Freshwater Shrimp Farming in Europe

Freshwater Shrimp Farming in Europe

Eco-responsible Practices for the Giant River Prawn

Géraud Laval

CABI is a trading name of CAB International

CABI
Nosworthy Way
Wallingford
Oxfordshire OX10 8DE
UK

Tel: +44 (0)1491 832111
E-mail: info@cabi.org
Website: www.cabi.org

CABI
200 Portland Street
Boston
MA 02114
USA

Tel: +1 (617)682-9015
E-mail: cabi-nao@cabi.org

Originally published in French under the title *Élevage de crevettes d'eau douce en Europe: Pratiques éco-responsables pour* Macrobrachium rosenbergii by Géraud Laval. © Éditions Quae. 2022

A catalogue record for this book is available from the British Library, London, UK.

ISBN-13: 9781836993148 (hardback)
         9781836993155 (ePDF)
         9781836993162 (ePub)

DOI: 10.1079/9781836993162.0000

Commissioning Editor: Jamie Lee
Editorial Assistant: Theresa Regueira
Production Editor: James Bishop

Translator: DeepL
Typeset by Straive, Pondicherry, India
Printed in the USA

Contents

Foreword

The large tropical freshwater shrimp, known as "Great Arms", *Macrobrachium* in Latin, *ouassous* ("king of springs") in the West Indies and chevrette in the Pacific, is native to south-east Asia, from India to Papua New Guinea. It has a unique life cycle: although it develops and reproduces in freshwater, the larval stage takes place in brackish water, which brings it into close contact with the large sea shrimps, its penaeid 'cousins', the stars of the world market. Several million tonnes of the latter are cultivated. The chevrette is more modest: less than a million tonnes, but its flavour is considered subtle by connoisseurs. It is to shrimp what crayfish is to langoustine: a delicacy of choice for chefs and gourmets alike.

Although its biological cycle had been completed by 1961, its rearing began on an experimental scale in ponds at the University of Hawaii, under the guidance of Takuji Fujimura, the father of the so-called "green water" method, as he first sought to reproduce the natural brackish water lagoon environments where the species' larvae are found. In the early 1970s, he succeeded in producing thousands of post-larvae, which he then seeded in earthen basins, well replenished with fresh water. Yields were low and irregular, but it worked. This enabled him to start spreading the technique in several tropical countries, including Thailand, Malaysia and Mauritius.

Shortly afterwards, the Centre national pour l'exploitation des océans (CNEXO) had an aquaculture experimentation centre built in Tahiti. The experimental approach was radically different: Jean-Michel Griessinger, the 'father' of the method, chose to control all the farming parameters to the maximum in order to increase yield and reliability. He is precise, rigorous and patient, and knows how to surround himself with colleagues who are as passionate as he is. I was lucky enough to join his team in 1976. Over the years, the "clear water" larval rearing method was perfected and became the most efficient and reliable in the world. And it still is. The cylindrical-conical larval rearing tanks reach 9 m^3, a record that still stands today. Over the following decades, it was exported to almost all the French overseas departments and territories, and spread to many countries, including Brazil. Methods of growing, harvesting and adding value to the product were also improved, thanks in particular to work carried out over 8 years in French Guiana. A collective work summarising all French knowledge of this type of farming (1991) brought this major targeted research programme to a close. In the French overseas territories, private farms initially developed, but then collapsed for a variety of reasons, including chlordecone pollution in the West Indies. Elsewhere, this type of farming is making headway almost everywhere in tropical countries, although it usually remains on a limited, or even small-scale, scale, which has probably shielded it from the pathologies generated by the intensification of large-scale farming, as has been observed in the case of various aquaculture species.

After holding a number of senior positions in aquaculture in the Mediterranean, in 2006 I was asked to create the strategic intelligence and forecasting function within the general management of the French Research Institute for Exploitation of the Sea (Ifremer). One day in 2016, a veterinary surgeon in the Gers region phoned me to ask if I could help him unblock the growth of *Macrobrachium* larvae in the experimental farm he had set up at his home... There was little silence from me on the phone. My last hatchery cycle as manager dates back to... 1983! After a few technical questions, I attempt a diagnosis over the phone and make a number of specific recommendations. A week later, Géraud Laval called me back: after implementing the recommended adjustments, the hatchery was doing well and the larvae were developing. It will then be successfully carried through to metamorphosis. It can now seed its first outdoor pond. In 2019, he will take a highly instructive trip to the Pointe-Noire Aquaculture Park in Guadeloupe, where he will be able to observe the larval culture and pond rearing of *ouassous* in a tropical environment.

Five years after his first production trials and remarkable progress in all phases of rearing, he has mastered the fine art of shrimp farming, both in the hatchery and during the grow-out period. With great deal of patience and creativity, he has been able to adapt tropical procedures to the variability of ecosystems in south-west France. Larval rearing is reliable and tank seeding can be programmed to within a few days. Grow-out yields mean that orders can be taken from an increasingly wide range of customers. Reputable restaurants are beginning to include this original shrimp on their menus.

This comprehensive, highly practical guide is designed to encourage the development of breeding for this original species, which offers a wide range of ways of adding value, from production systems to marketing.

Tasting this French-bred species today can be described as a renaissance!

I'd like to thank Géraud Laval for his intuition about the feasibility of this type of breeding in mainland France, for his tenacity and creativity in developing it in a temperate climate, and for sharing what he's learned with all those who are interested in this wonderful adventure!

Denis Lacroix, Head of Foresight at Ifremer

Legend

This guide uses a system of internal cross-references based on numbered dots, from 1 to 24: the numbers in the body of the text or in the figures refer to calls for the same numbers, whose darker markings, visible through an underline, are in the margin of the text.

1. The different morphotypes of males.
2. Hierarchical social organisation.
3. Production in temperate zones in the United States.
4. EC Regulation no. 708/2007 on the use of alien species in aquaculture.
5. Larval rearing in hatcheries.
6. Types of nursery tanks.
7. Development of grow-out ponds.
8. Risk of shrimp escaping into the natural environment.
9. Placement of post-larvae in nursery.
10. Number of postlarvae to be introduced.
11. Installation of substrates in nurseries.
12. Daily management and water quality parameters.
13. Feeding post-larvae in nurseries.
14. Monitoring and optimising the growth of post-larvae in nurseries.
15. Preparation of basins.
16. Introduction of juveniles.
17. Day-to-day water management.
18. Phytoplankton management.
19. Feeding and fertilisation.
20. Growth monitoring.
21. Predator prevention.
22. Shrimp harvesting.
23. Live storage of shrimp after harvesting.
24. Slaughter and cold storage.

Acknowledgements

The author would like to thank all those who contributed to the production of this book, in particular Ségolène Calvez, Denis Lacroix, Jean-Marie Peignon, François Herman and Pierre Boudry for their meticulous proofreading and the relevance of their contributions. Special thanks to students Martin Quéro and Max Guézou for their contribution to this work.

Credits

Figures 1.2, 1.4, 1.7, 6.1 © Ifremer

Figure 1.3. Martin Quéro

Figures 1.5.a, 1.5.b, 2.4, 2.5, 2.7, 2.9, 2.11, 3.2, 3.4, 4.2, 4.3, 4.4, 4.6, 6.2, 6.3, 6.4, 6.6, 6.8 © Géraud Laval

Figure 2.2© François Herman

Figure 2.6. © Marc Zalio

Figure 2.8. © Association for the Promotion of Fish from the Dombes Ponds (APPED)

Figure 2.10. © Maria Ruiz Bascaran

Figures 2.12, 3.3, 4.5, 6.5, 6.7 © Gascogne Aquaculture

Figure 5.1. © Martin Quéro/Gascogne Aquaculture

Introduction

This guide to good farming practice for the tropical freshwater shrimp *Macrobrachium rosenbergii* in temperate zones is intended primarily for aquaculturists and other aquaculture professionals. It may also be useful to students, teachers and researchers in the fields of aquaculture, animal husbandry or environmental sciences, as well as to private individuals wishing to carry out a project or satisfy their curiosity. Its aim is to teach techniques for designing and managing semi-extensive tropical freshwater shrimp (*Macrobrachium rosenbergii*) farming units in a southern European climate. Freshwater shrimp farming for consumption is currently non-existent in mainland France and the European Union (excluding the French overseas departments). In Guadeloupe, a production site located at Pointe-Noire (Parc Aquacole) produces this shrimp, known locally as *ouassous*. These farms are widespread in tropical and subtropical areas, mainly using semi-intensive production methods in Asia.

This guide is based on initial production experiments in France by Gascogne Aquaculture between 2017 and 2021 in the Gers, in collaboration with ONIRIS-INRAE. It often refers to *Freshwater prawns - biology and farming*, by M. B. New *et al* (2010), a reference work for this species. This guide is a first edition that can be enriched in the years to come by the experience of other pioneering producers who will embark on the adventure of freshwater shrimp farming in temperate zones using this work.

i.1. Rethinking tomorrow's aquaculture practices

This guide is in line with the environmental and societal challenges facing aquaculture at the start of the [21st] century. It aims to contribute to Europe's self-sufficiency in aquaculture products and to promote eco-responsible farming practices.

Our era is marked by unprecedented climatic, environmental and societal events. The global pandemic caused by Covid-19 has highlighted the importance of local food autonomy, calling into question our dependence on imported food. Many consumers prefer local products. The saturation of means of transport and/or the closure of borders caused by the confinements of early 2020 have made European consumers aware of the fragility of world trade and the need to acquire or maintain our food sovereignty. Livestock sectors dependent on imported inputs are, in fact, fragile, and their environmental cost is forcing us to reduce them.

Public authorities in Western Europe are now encouraging us to rethink our production and consumption patterns in all areas, including food.

This manual and the farming practices detailed in it take these factors into account and offer a concrete example of an innovative, virtuous and local production model for exotic shrimps in France. France imports over 99% of the shrimp it consumes. Imported shrimp is produced by industrial methods that are all too often disrespectful of the environment (pollution, destruction of mangroves, depletion of fish stocks, negative carbon footprint, etc.). However, it is possible to produce some of these species locally using simple, low-input techniques, as proposed in this book with *Macrobrachium rosenbergii*.

This species is adapted to warm waters, and is therefore prepared for the predicted rise in temperatures over the coming years; its yield would even be improved by warmer, longer summers in our latitudes. The proposed low-input system does not require the input of marine proteins (fish meal and fish oil), a major input in intensive aquaculture, which is drawn from natural resources. It consumes little fossil energy and little water (natural heating of the breeding ponds in summer, little or no water renewal). It can discharge no effluent into the natural environment, thanks to recycling in closed installations. These virtuous criteria meet the growing expectations of consumers who are looking for fresh, local products produced in an eco-responsible way.

Finally, society has growing expectations in terms of quality, health and animal welfare. Farmed at a very low density (3 or 4/m^2), the shrimps are less stressed and do not need to be treated (no antibiotics). Sold fresh and on the spot, they do not require the addition of preservatives (sulphites) and the excellence of their natural flavour guarantees their commercial success with gourmet food lovers, as many do in Europe.

The proposed *M. rosenbergii* farming model is robust, sustainable and resilient. Local production using exclusively local feed inputs and juveniles produced in France, combined with the low level of technical skill required, means that production can be self-sufficient.

i.2. Contents

This guide focuses mainly on the pre-growth (nursery) and grow-out phases in outdoor ponds, and only briefly covers the earlier phases in hatcheries.

After a first chapter presenting the species (characteristics and biology), the second chapter looks at the main principles of its farming in temperate zones. The European regulatory context relating to the introduction of an exotic aquaculture species, the production cycle and the infrastructures required for its rearing are detailed.

Chapters 3 and 4 are more operational in nature: they describe in detail the day-to-day operations involved in farm management (tank preparation, growth monitoring, feeding, etc.), for the nursery and grow-out phases respectively. To facilitate practical use in real-life rearing situations, these two parts are written in the form of thematic sheets, structured in a uniform way and covering the following points: objective, specificity of the species or system, points to watch, indicators of good practice, instructions for implementation, recommendations and special precautions.

Chapter 5 looks at the main diseases and other health problems that can be encountered in nurseries or grow-out facilities.

Finally, Chapter 6 deals with the final harvesting and marketing of shrimp with a view to supplying fresh produce locally.

Macrobrachium rosenbergii: Characteristics and Biology

1

1.1. Systematic position

The freshwater shrimp *Macrobrachium rosenbergii* is a crustacean of the order Decapoda, which has 5 pairs of legs and includes the vast majority of edible crustaceans: prawns, crayfish, lobsters, crabs, etc.

The classification distinguishes Carid shrimp, which includes *Macrobrachium rosenbergii*, from marine shrimp of the Peneid family, which includes the world's most important farmed shrimp species (Figure 1.1). The two main species farmed are the white-legged shrimp (*Penaeus vannamei*) and the tropical tiger shrimp (*Penaeus monodon*). The kuruma or imperial shrimp (*Penaeus japonicus*) is farmed on the French Atlantic coast. The distinguishing features of the Caridae are mainly the type of gills, the absence of claws for the 3rd pair of legs and a more advanced larval stage at birth: zoea rather than nauplius.

Macrobrachium belongs to the Palaemonidae family, which also includes a shrimp that is well known on Europe's Atlantic coasts, *Palaemon serratus*, known as the "bouquet shrimp".

The genus Macrobrachium is characterised by hepatic spines and the absence of supraorbital spines and gills, by the last 3 pairs of legs without pincers and a large, serrated rostrum.

Around 125 species have been described in the *Macrobrachium* genus, of which a dozen species are bred or have been the subject of experiments, with *Macrobrachium rosenbergii* being the most widespread. In French, it is known as "bouquet géant" (FAO), "chevrette" (Tahiti, New Caledonia, French Guiana) or "ouassou" (Guadeloupe). The commercial names authorised in France (DGCCRF) for *M. rosenbergii* are "chevrette" or "crevette tropicale d'eau douce".

DOI: 10.1079/9781836993162.0001

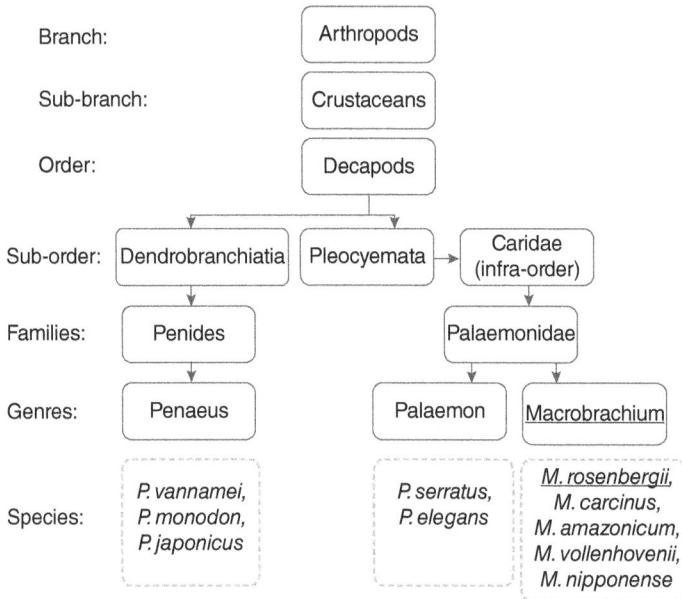

Figure 1.1. Simplified classification of shrimps.

1.2. Distribution

1.2.1. Natural distribution

The original natural distribution of *Macrobrachium rosenbergii* extends from Papua New Guinea in the east to Pakistan in the west (Figure 1.2). It includes northern Australia, Indonesia, the Philippines, Malaysia, southern China, South-East Asia and southern India.

In these areas, shrimp are found in warm (19 to 32°C) and shallow (1 to 2 m) waters, in fresh or brackish water, in lakes, rivers, ponds and estuaries. Brackish water is essential to complete the reproductive cycle.

The introduction of Macrobrachium *rosenbergii* for aquaculture, in regions where it was absent and where climatic conditions were favourable, has led to the extension of its area of distribution in the natural environment. *M.* rosenbergii has become acclimatised in Brazil, French Guiana, Venezuela and Ecuador, without any assessment of its impact on local ecosystems.

1.2.2. Distribution of livestock production

Figure 1.3 shows *Macrobrachium rosenbergii* production in tonnes per year by country. The top producing countries are in South-East Asia, with China producing 140,000 tonnes in 2019, followed by India, Bangladesh and Thailand.

Figure 1.2. Map of the natural distribution of *Macrobrachium rosenbergii* (source: Ifremer, 1991).

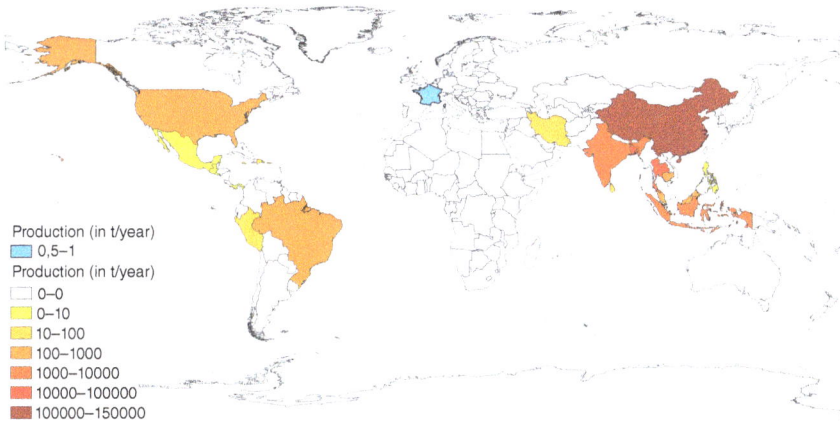

Figure 1.3. World map of *Macrobrachium rosenbergii* production, based on FAO data (2019), excluding France.

Global production in 2019 was 273,000 tonnes (FAO). By comparison, global production of farmed shrimp of all species will exceed 5 million tonnes in 2019, with around 80% of this coming from the *Penaeus vannamei* species alone.

1.3. Morphological characteristics

1.3.1. External morphology

The *Macrobrachium rosenbergii* shrimp has a life cycle divided into several stages: larval, post-larval (or juvenile) and adult.

The female incubates her fertilised eggs under the abdomen for around twenty days (Figure 1.5.a). The eggs are orange when laid, turning brown and then grey at the end of incubation. When they hatch, the larvae are at the "zoe" stage, measuring barely 2 mm.

The larva differs greatly morphologically from the adult. It is a free-swimming planktonic organism whose head (cephalon) has a pair of eyes and functional appendages. Its larval development is punctuated by 11 successive moults, resulting in an evolution of its morphology and 11 differentiated stages. The passage from stage XI to stage XII leads to a metamorphosis of the larva into a post-larva. The post-larva measures around 8 mm and weighs between 6 and 10 mg in the first few days. Its morphological characteristics are those of an adult, with the exception of its undeveloped sexual organs.

Figure 1.4 shows the external morphology of the adult. They can reach a size of 32 cm, excluding claws, for males, and 25 cm for females. The record is a male weighing 600 g (Ifremer, 1991). Their body, which is greenish to greyish brown with touches of blue, is made up of 20 segments called somites. Each of these somites has a pair of appendages with specific names and functions, as described in Table 1.

The head (6 somites) and thorax (8 somites) are fused together to form the **cephalothorax**. This alone accounts for 55-60% of the animal's weight. It has 5 pairs of legs, the pereiopods, the first two of which, also called chelipeds, have pincers.

The **abdomen**, made up of 6 somites, ends in the telson. The appendages, the pleopods, ensure swimming in open water. In the female, the lateral walls of the first 3 abdominal segments are elongated downwards to form a chamber in which the eggs are incubated.

M. rosenbergii can be recognised by its particularly long rostrum. It has a large number of teeth: 8 to 15 on the dorsal side and between 6 and 16 on the ventral side. The eyes are located on either side of the rostrum and are attached *via* a peduncle to the first somite: the acron.

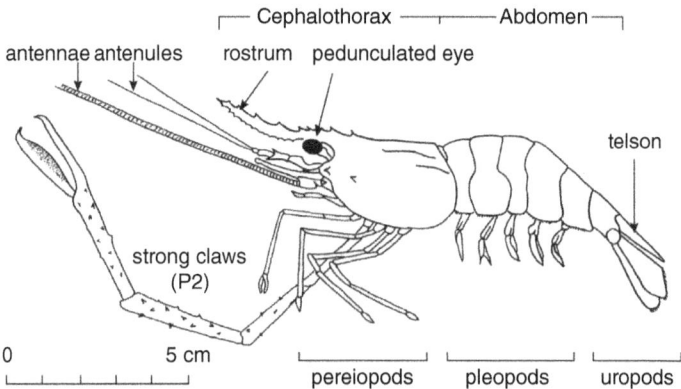

Figure 1.4. External morphology of *Macrobrachium rosenbergii* in the adult stage.

Table 1. Names and functions of the appendages on the somites of *Macrobrachium rosenbergii* (source: New (FAO), 2002).

Body section	Somite	Name of appendices	Function of the appendages
Head of the cephalothorax	1	Embryonic segment (not visible in adults)	
	2	1st antennas	Sensory and tactile perception
	3	2nd antennas	Tactile perception
	4	Mandibles	Food grinding
	5	1st maxillae (maxillule)	Food intake
	6	2nd maxillae	Food intake and water circulation thanks to scaphognathite
Thorax part of the cephalothorax	7	1st maxilliped	Food intake
	8	2nd maxilliped	Food intake
	9	3rd maxilliped	Food intake
	10	1st pereiopod (1st cheliped)	Hunting
	11	2nd pereiopod (2nd cheliped)	Hunting and reproduction
	12	3rd pereiopod	Mobility (walking); gonophores on the base of the appendages for females
	13	4th pereiopod	Mobility (walking)
	14	5th pereiopod	Mobility (walking); gonophores on the base of the appendages for males
Abdomen	15	1st pleopod	Mobility (swimming)
	16	2nd pleopod	Mobility (swimming); reproduction for males
	17	3rd pleopod	Mobility (swimming)
	18	4th pleopod	Mobility (swimming)
	19	5th pleopod	Mobility (swimming)
	20	Telson or uropod	Mobility (swimming); excretion

M. rosenbergii can also be identified by its two pairs of chelipeds, legs with pincers. The second pair of chelipeds is very long in this species, particularly in males, giving the genus its name (*Macrobrachium*). The last 3 pereiopods are motor legs and have no claws. There is also a hepatic spine on the side of the cephalothorax and no supraorbital spines or gill rakers.

1.3.2. Respiratory and circulatory systems

M. rosenbergii breathes thanks to gills located in the gill chamber and protected by the branchiostegite (lateral flap of the carapace). These gills are supplied with water by a current created by the scaphognathite. The water enters through the space between the branchiostegite and the appendages, crosses the gill chamber and leaves anteriorly. Gas exchange in the gill takes place by diffusion, facilitated by this forced ventilation. This allows oxygen to be added to the haemolymph as it passes through the gills. The haemolymph then reaches the heart, located behind the cephalothorax, and is sent to the various tissues via a complex arterial system. The oxygen-depleted haemolymph returns to the gills via collecting ducts and sinuses.

1.3.3. Digestive system

The digestive system begins at the head with the mouth and mouth parts and ends with the anus at the base of the telson. A short oesophagus leads to a masticatory stomach or "proventriculus", which is divided into two chambers. The muscular anterior chamber, or "cardiac stomach", has the role of kneading the food to reduce it to small particles that can pass through the filter of the posterior chamber, or "pyloric stomach". The middle intestine (mesentery) that follows absorbs nutrients; it is connected to an important digestive gland in crustaceans, the hepatopancreas. The hepatopancreas occupies a large part of the cephalothorax and is responsible for secreting digestive enzymes and storing energy and mineral reserves. Indigestible material is transformed into faecal matter in the midgut, which then passes into the relatively short hindgut before being released through the anus.

1.3.4. Reproductive system and sexual dimorphism

The gonads are located inside the cephalothorax. In females, the ovaries are located dorsally and give rise to a pair of oviducts that terminate in the genital orifices (gonopores) located between the pereiopods. The male reproductive system consists of a pair of fused testes and vas deferens ending in terminal ampullae containing spermatophores and opening onto the gonopores. Shrimp can be sexed by observing the location of the gonopores. These are located on the 3rd pair of pereiopods for the female and on the 5th pair for the male.

Females are generally smaller than males. Mature females also have proportionately smaller heads and claws than males (Figure 1.5.a). Males can be recognised by their distinctive morphology (Figure 1.5.b). ❶ There are 3 male morphotypes: large males with blue claws, males with orange claws and small males with light-coloured claws. These morphotypes are characterised by the size of the individual, as well as the size and colour of the claws, resulting in a hierarchy in population ❷. Males with blue claws have the longest claws (one and a half times the size of the body) and are the dominant males. Then there are the orange-clawed males, followed by the smaller males with the smallest claws.

1.4. Biology

1.4.1. Life cycle and reproduction

Macrobrachium rosenbergii has a catadromous life cycle (Figure 1.6). Juveniles and adults live in tropical freshwater and part of the cycle, the larval phase, takes place in brackish water with a salinity of between 12 and 16 psu (or g/l).

The larval stage, which takes place in brackish water, lasts around 1 month and consists of a succession of moults before metamorphosis. During this month, the larvae are pelagic and live in open water, swimming upside down with their ventral side towards the surface and their telson in front. They will

(a)

(b)

Figure 1.5.a. Female of *M. rosenbergii* with seeds.
Figure 1.5.b. Two morphotypes of *M. rosenbergii* males: blue claws in the foreground and orange claws in the background.

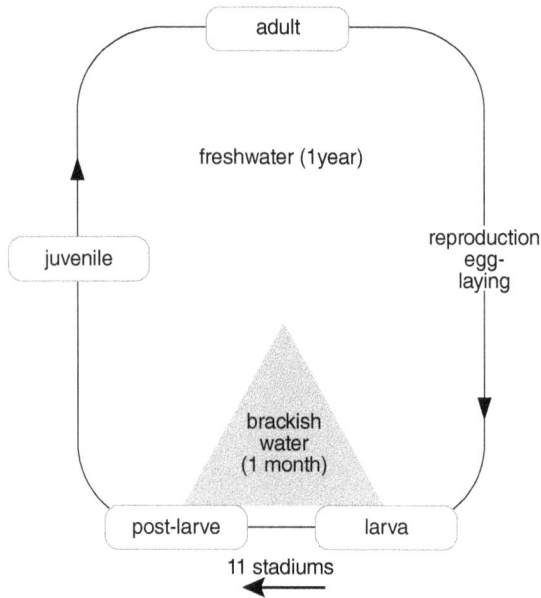

Figure 1.6. Diagram of the catadromous life cycle of *Macrobrachium rosenbergii*.

only be morphologically similar to adults after metamorphosis, and will then be called post-larvae, or juveniles.

Post-larvae are benthic and migrate upstream to freshwater, where they spend most of the rest of their lives.

After around 5 months, the shrimp are sexually mature. In the natural environment, reproduction can be seasonal or continuous, depending on geographical location and temperature.

Mating takes place following the pre-mating moult of a female; the male, most often a dominant blue-clawed male, deposits a spermatophore in the female's seminal receptacle, located near her genital opening. Eggs are laid within 24 hours of mating. The eggs are fertilised as they pass through the spermatophore. Females lay between 5,000 and 20,000 eggs the first time they lay, and can lay up to 100,000 eggs when they are larger; the number of eggs is around 7,500 per 10g of female body weight. Gilled females keep their eggs under the abdomen throughout embryonic development, which lasts around twenty days. The larvae then hatch and passively drift down the river towards a brackish estuary where they can develop.

1.4.2. Diet and nutritional requirements

The larvae are carnivorous and feed on zooplankton. The ability of the larvae to ingest and digest changes with each larval stage. From stage I to stage V, the larvae have a small hepatopancreas and do not produce enough enzymes to digest a multitude of prey. They feed solely on small zooplankton. As the hepatopancreas develops, the diet diversifies to include larger zooplankton, small worms and larvae of other crustaceans.

After metamorphosis into post-larvae, the shrimp's diet changes and becomes omnivorous. It then consists of a mixture of algae, aquatic plants, molluscs, insects, worms and other small crustaceans. Post-larvae need a lower protein diet than larvae. They grow best when their diet contains 35% or more protein.

For post-larvae and adults, low lipid levels of 2% are sufficient if the lipids contain the necessary essential fatty acids and the protein intake is adequate. The lipid level suggested by the FAO (New, 2002) is 5%. Some polyunsaturated fatty acids cannot be synthesised, such as linoleic acid - omega 6 type - and must be provided by food. They are also unable to produce sterols, so a cholesterol intake of around 0.5 to 0.6% is essential. According to some authors, a fat content of more than 10% can lead to reduced growth and is not recommended (New *et al.*, 2010).

Carbohydrates are not essential for this species, but can be used as a source of energy. In particular, *M. rosenbergii* is capable of assimilating complex carbohydrates such as starch and cellulose (New *et al.*, 2010). Finally, mineral requirements have been identified, such as zinc (90 mg/kg), calcium and phosphorus, in a Ca/P ratio of the order of 1.5 to 2 to 1 (New, 2002).

Adults have very similar nutritional requirements to post-larvae. They just need less protein, from 3 months onwards, at around 30-35%.

In extensive and semi-extensive farming (less than 3.5 shrimp/m²), feed is provided by the natural productivity of the pond, if it is sufficiently fertile. Fertilisers, preferably organic, are recommended. The ecosystem is then capable of producing enough plant or animal organisms (zooplankton, invertebrates) to cover the nutritional requirements of the shrimps, particularly in terms of proteins and essential fatty acids. For higher densities, an industrial compound feed should be provided.

Figure 1.7 illustrates the place of freshwater shrimp (chevrette) in the trophic chain of a farmed pond. The pellets provided are one of the shrimp's food sources, which is predominant at high densities.

The advantage of low-density systems is that they allow shrimps to be fed naturally and therefore avoid the need to supplement with marine proteins and oils, rich in omega 3 and 6, taken from the natural environment (forage fish), which are increasingly rare and expensive. This is a perfect response to the challenges of agro-ecology, although the yields obtained per unit area are lower.

1.4.3. Growth

A feature common to all arthropods, including crustaceans such as *M. rosenbergii*, is their discontinuous growth through successive moults. At regular intervals, the shrimp sheds its exoskeleton, its carapace; this is called exuviation. This

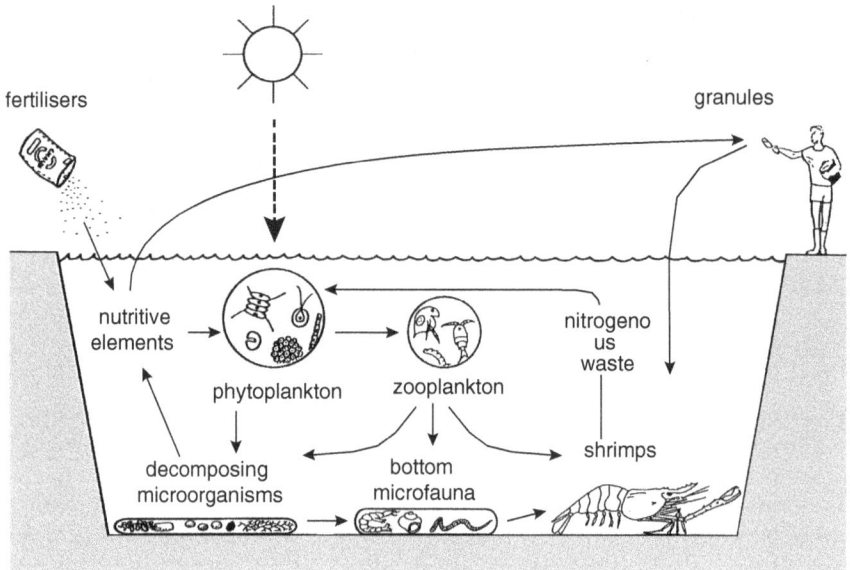

Figure 1.7. Trophic chain in a *M. rosenbergii* rearing pond.

moult allows the shrimp to grow by extending its body *through* a massive ingress of water, followed by the reconstitution of a new rigid carapace by mineralisation over a period of a few hours. The inter-molt period that follows lasts from a few days to a few weeks; this is a phase of accumulating reserves in preparation for the next molt.

The moult and the hours that follow are a period of great vulnerability for the shrimp, as its shell is soft. It is a period of high energy and oxygen consumption. In particular, the shrimp is exposed to predation and cannibalism; its sensitivity to toxic substances and diseases is also increased. In *M. rosenbergii*, copulation takes place after moulting, so the female is protected by the male. High stocking densities increase the risk of succumbing to cannibalism, latent toxicity or anoxia.

1.4.4. Hierarchical social organisation ②

Macrobrachium rosenbergii shrimps have a well-defined social organisation. This is established through agonistic behaviour and is reflected in the morphotypes of the males. There are three morphotypes: blue claws (PB), orange claws (PO) and small light claws (PC), each with its own specific behaviour.

Males differ not only in the length of their claws but also in size ①. Males with blue claws are aggressive, dominant and very territorial. Orange-clawed males are sub-dominant and non-territorial. Small light clawed males (PC) are submissive, highly mobile and non-territorial; they grow slowly. Each male can evolve and move up the hierarchy, so a PC becomes a PO and a PO becomes a PB if he takes the place of a PB on the territory. The PB morphotype is at the top of the social structure; it is the irreversible and sexually active final stage. The PO morphotype, on the other hand, puts its energy into growth; it has a highly developed digestive system and a very reduced reproductive system.

The hierarchy in the population is established through fights and parries. The size of the claws and the individual plays a major role in the outcome of these interactions, explaining the dominance of the blue-clawed males.

The distribution of morphotypes in a population is approximately 50% small light clawed males (PC), 40% orange clawed males (PO) and 10% large blue clawed males (PB). However, the proportion of small males (PC) increases with density, unlike the other two morphotypes, PO and PB.

The establishment of a hierarchy affects the development of shrimp. In a population, dominant and aggressive individuals eat more and grow faster at the expense of dominated individuals. In high-density systems, a drop in average population growth is observed.

These limitations are particularly important in the context of temperate zone farming, which is characterised by a short production season. Consumers are looking for large individuals, so low-density systems will be favoured because they will produce individuals of a size better suited to this market in a short space of time.

Principles of Breeding in Temperate Zones

2

2.1. Climate preferences and limits

The freshwater shrimp *Macrobrachium rosenbergii* has been farmed in its natural tropical area of origin in South-East Asia for a very long time. Originally, small-scale farming was carried out using wild juveniles taken from rivers. In the early 1960s, breeding took off with the development of larval culture in brackish water by Shao-Wen Ling in Malaysia. Following its introduction in Hawaii in 1965, commercial breeding developed in the United States thanks to the research work of Takuji Fujimura at the University of Hawaii.

The choice of rearing techniques for *Macrobrachium rosenbergii* depends on climatic and environmental conditions, the economic context (availability of inputs, labour costs, profitability, existence of a market) and local regulations. Societal expectations, particularly with regard to environmental impact and animal welfare, are playing an increasingly important role in Western countries, across all livestock sectors.

In the tropics, production takes place in a continuous system, with a long rearing cycle (over more than a year), regular reseeding and regular selective harvesting.

In temperate zones, year-round production of *M. rosenbergii* is not possible because of the winter cold, the lethal temperature for this species being below 13°C. This avoids the risk of invading natural environments, which has become a major concern over time with the introduction of exotic species. This climatic context and these constraints have led to the development of a discontinuous rearing system, including a preliminary pre-pregnancy phase.

This discontinuous system is characterised by a single annual production run over a short period in outdoor ponds, and by total harvesting by emptying the pond at the end of the cycle before the cold arrives.

In temperate zones, ponds cannot be used for shrimp farming during the cold period, which lasts 7 to 8 months. Undesirable and predatory species can

DOI: 10.1079/9781836993162.0002

develop during this period, and this must be taken into account. However, the ponds can be used advantageously during the winter period to carry out part of the production cycle of another cold-water aquaculture species, such as trout.

The seasonal nature of the harvest can be seen as a constraint in temperate zones. Harvesting can only take place in early autumn, and marketing of the live or fresh product is spread over a short period of just a few weeks.

However, cultivation of *M. rosenbergii* has an essential advantage. Unlike marine species (peneidae), freshwater shrimp can be produced in continental areas, far from coastal zones. This makes it possible to set up production sites close to large inland urban markets, and to supply them with high-quality fresh shellfish.

2.2. Production in temperate zones in the United States ●

A model for rearing *M. rosenbergii* in temperate zones was described in the United States in the early 2000s by the University of Kentucky, thanks to the research work carried out by James H. Tidwell *et al* (2002).

In the temperate zones of the eastern United States, the outdoor growing period is short, from 100 to 140 days a year, during the summer. This favourable period for the growth of *M. rosenbergii* is framed by colder, more sensitive transition periods (late spring and early autumn). In spring, juveniles can be introduced into outdoor ponds when minimum water temperatures (at the end of the night) have reached 19-20°C and are unlikely to fall again over the following days. At the end of the period, early autumn cold can cause significant losses.

Because of these limitations, the temperate zone breeding model has its own constraints and adapted procedures must be respected.

Juveniles introduced into outdoor ponds must be between 45 and 60 days old (weighing between 0.2 and 0.5 g) and are therefore pre-grown in the nursery. This stage enables juveniles that have already reached an advanced stage of growth to be placed in ponds, thereby compensating for the short grow-out. Compared with young post-larvae, these pre-fattened juveniles are more robust and have a better chance of survival, particularly with regard to predation.

Low stocking densities for juveniles are to be preferred because individual growth of *M. rosenbergii* is dependent on the density of the individuals. High densities encourage early sexual maturation, which limits growth. At low densities, competition between individuals and sexual maturation are delayed, resulting in better individual growth.

In the United States, two cultivation techniques are described:

- *Low-input* or semi-extensive cultivation, with an initial stocking density of between 2 and 3.5 juveniles/m^2.
- *High technology* or intensive cultivation, with a stocking density of 6 to 7.5 juveniles/m^2.

In low-input cultivation, plant feed is sufficient to fertilise the water in the ponds, with a low protein content (17-20%). Production yields can reach 800 to 1,000 kg/ha.

In intensive cultivation, protein-rich feed must be provided, the ponds must be equipped with substrate and aerated, and the juveniles are sorted before introduction. Production yields can reach over 2,000 kg/ha.

Under optimum conditions in tropical zones, yields of 3.5 tonnes per hectare per year have been observed, notably in Martinique in the 1980s (personal communication, Denis Lacroix).

2.3. Regulatory conditions for its introduction into Europe (EU)

In Europe (excluding overseas territories) the first trial of outdoor production of *M. rosenbergii* was carried out in France in 2017 by Gascogne Aquaculture in the Gers, in partnership with ONIRIS-INRAE. In 2018, production for commercial purposes was authorised and is currently ongoing.

Community Regulation No. 708/2007 ❹ of 11 June 2007 (amended by Regulation 304/2011 of 9 March 2011) concerning use of alien and locally absent species in aquaculture lays down the rules for the introduction of a new aquaculture species, known as "alien", into the European Union.

An **exotic species** is defined in Regulation 708/2007 as follows:

a)   any species or subspecies of aquatic organism present outside its known natural range or its natural range of potential dispersal;
b)   all polyploid organisms and fertile species obtained by hybridisation, whatever their natural range or potential dispersal.

Annex IV lists exotic species for which the Regulation does not apply, in other words exotic species authorised to be farmed in the European Union (in open aquaculture facilities). *M. rosenbergii* is only included on this list for the French overseas departments.

Accordingly, the *M. rosenbergii* species cannot be reared in open aquaculture facilities (i.e. facilities that are not separated from the wild aquatic environment) in Europe (excluding overseas territories). Its introduction would require the submission of a permit application to the competent authority of the Member State of destination, the Ministry of the Environment in France, together with an environmental risk assessment.

However, this complex procedure can be avoided if the receiving farm is recognised as a **closed aquaculture facility**.

The definition of a "closed aquaculture facility" within the meaning of Regulation 708/2007 is *a land-based facility*: a) in which : - aquaculture is carried out in an aquatic environment involving recirculation of water ; - discharges have no connection whatsoever with open waters prior to screening and filtering or percolation and treatment to prevent the release of solid waste into the aquatic environment and any escape from the facility of farmed species and non-target species likely to survive and subsequently reproduce ; b) and which : - prevents losses of farmed or non-target species and other biological material, including pathogens, due to factors such as predators (e.g. birds) and flooding (e.g. the facility must be located at a safe distance from open water following an appropriate assessment by the competent authorities); - prevents, by reasonable means, losses of farmed or non-target species and other biological material, including pathogens, due to theft and vandalism; - ensures the proper disposal dead organisms.

Broadly speaking, a closed aquaculture farm is designed so that its rearing water cannot be discharged into natural waters. A water recirculation system between rearing tanks, or a filtration system recognised by the government, must be implemented.

The regulatory procedure chosen by SARL Gascogne Aquaculture is recognition by the French authorities as a closed aquaculture facility. In 2021, other farms are in the process of obtaining this recognition, paving the way for *M. rosenbergii* farming in France, subject to certain conditions.

This guide to good practice and the recommendations for setting up a rearing site for *M. rosenbergii* fall within the regulatory framework of a closed aquaculture operation.

In addition to the regulatory aspects, the closed facility offers a number of advantages for breeding *M. rosenbergii*:

- It limits the risk of introducing harmful or predatory wild species into the rearing tanks.
- It uses less natural water, helping to preserve the natural aquatic environment.
- It avoids the risk of chemical contamination (pesticides, etc.) and biological contamination (pathogenic micro-organisms) that can be fatal to farmed species.
- It helps to maintain the disease-free status of farmed species.

The status of closed aquaculture facility is recognised by the relevant departments of the Prefecture of the département in which the farm is located, i.e. the Direction départementale de la protection des populations (DDPP) and the Direction départementale des territoires (DDT). For a new fish farm, the application **to create a water body** must be submitted to the DDT.

Other regulatory procedures, not specific to the introduction of an exotic species, apply:

- declaration file under the Water Act to the DDT for a cumulative surface area of ponds between 0.1 and 3 ha, or authorisation file if it is greater than 3 ha;
- obtaining zoo-sanitary approval for marketing from the DDPP;
- ICPE (Installation Classée pour la Protection de l'Environnement) authorisation file with the DDPP if the planned production exceeds 20 t/year;
- declaration of borehole creation to the DDT (where applicable).

Other declarations or authorisations may be required under regulations specific to protected areas if the farm is located in an area classified as a Natura 2000 site, a wetland, etc. The DDT will be able to provide you with information.

2.4. The stages in the production cycle

The production cycle is divided into three distinct stages: larval rearing in hatcheries, pre-growth in nurseries and grow-out in outdoor ponds (Figure 2.1).

2.4.1. Larval rearing in hatcheries ⑤

Hatchery larval rearing is the first stage in the *Macrobrachium rosenbergii* production cycle. Females are fertilised by sexually active "blue-clawed" males from breeding stock (or from the wild in tropical areas of origin). After 3 weeks of incubation, the seeded females release their larvae, at a rate of 10,000 to 50,000 larvae per shrimp depending on their weight. The larval culture that follows is a delicate phase requiring dedicated infrastructures. The larvae are reared in brackish water for 20 to 30 days. These may be open-flow systems in tropical zones or recirculating systems requiring mechanical and biological filters (Figure 2.2). At the end of the cycle, after 11 larval stages, the larvae metamorphose into post-larvae (PLs). The post-larvae are acclimatised to freshwater within a few hours and can then be transferred to a nursery. In tropical zones, it is possible to release young PLs a few days old directly into the grow-out ponds, as there are no climatic limits.

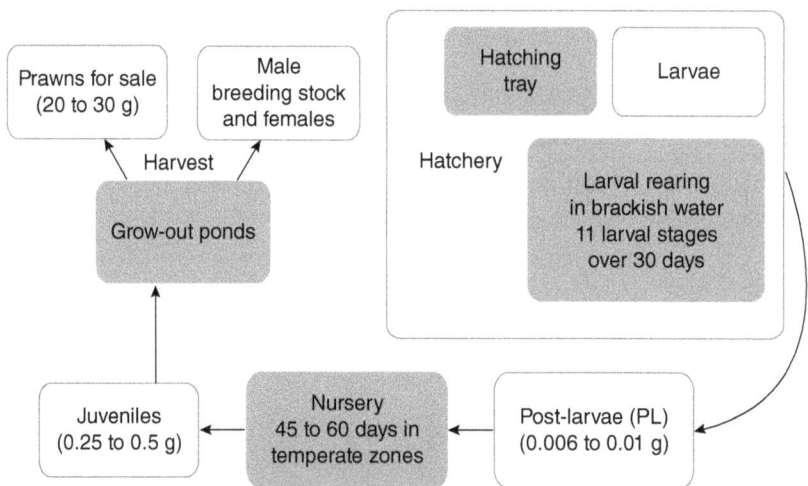

Figure 2.1. Production cycle of *Macrobrachium rosenbergii*.

Figure 2.2. *Macrobrachium rosenbergii* hatchery equipped with 7 m³ cylindro-conical tanks (Parc aquacole - OCEAN SAS, 97116 Pointe-Noire, Guadeloupe).

2.4.2. Pre-pregnancy in nurseries

The pre-growth stage in a nursery is necessary in temperate zones because of the limited duration of outdoor grow-out. It enables the post-larvae to start growing before being placed in grow-out ponds in late spring. This pre-growth is carried out in above-ground rearing tanks in a recirculating system (RAS[1]); it lasts 45 to 60 days and produces juveniles[2] weighing between 0.2 and 0.5 grams.

Pre-fattening lengthens the production cycle, which would be too short to achieve a marketable size if it were limited to the summer period in out-door ponds in a temperate environment. The advanced size of juveniles helps to limit losses due to predation when they are introduced outdoors.

The nursery only operates during the 2 months before the fish are put into the grow-out pond. In the south of France, this corresponds to the period from the end of March to the end of May.

The temperate zone nursery is a building (or greenhouse) housing pre-growth tanks of a size adapted to the number of PLs. These tanks contain fresh water heated to between 25 and 28°C.

Chapter 3 of this guide is devoted to nursery management.

2.4.3. Grow-out in outdoor ponds

Grow-out in ponds is the final stage in the rearing cycle. Ponds with a surface area ranging from 1,000 to 10,000 m² and shallow depths (1 metre) are used.

Juveniles from the nursery are brought in as soon as temperatures allow (above 19-20°C) and will grow until the final harvest. This period lasts from 3 and a half to 5 months in temperate zones (between mid-May and mid-October) and produces shrimps weighing between 20 and 30 grams, or even more if the technique is mastered and the climatic conditions are favourable.

Chapter 4 of the book is devoted to the management of nursery ponds.

The infrastructure and equipment required for a nursery and grow-out ponds are detailed below.

2.5. Livestock infrastructure

This guide is limited to a description of the infrastructures and rearing practices for the nursery and grow-out stages. The hatchery stage requires considerable technical expertise and is only briefly discussed ⑤. In shrimp production, but also for other animal species, this stage is usually handled by a specialised workshop that can distribute the juveniles (or post-larvae for shrimp) to a large number of grow-out farmers. In France, Gascogne Aquaculture (Gers) offers this service for mainland France, and the Parc Aquacole de Pointe-Noire (Guadeloupe) for the West Indies.

2.5.1. The nursery

The nursery will be designed to accommodate the number of postlarvae required to seed the farm's grow-out ponds. It can also be designed to accommodate postlarvae from several farms, allowing fish farmers who wish to concentrate on managing their grow-out ponds to dispense with the nursery by carrying out this stage in a collective structure.

2.5.1.1. Recirculation system (RAS)

The technique used in nurseries in temperate zones is a closed recirculating "RAS" system. This system allows continuous reuse of water through bio-filtration, reducing ammonia toxicity through nitrification.

Figure 2.3 shows the main elements of the system.

2.5.1.2. Sizing of breeding tanks

The volume of the nursery tanks will be calculated according to the rearing objectives and therefore the number of juveniles to be produced.

An average of 2,000 PLs/m^3 is used to size the infrastructure, provided that the shrimp are supported (substrates). It is the density per unit area that determines the number of shrimp per tank ⑩.

Table 2 shows the example of a farm wishing to seed 1 ha of grow-out ponds. The introduction of 40,000 PLs in nurseries is suggested, taking into account mortality (of around 20%) during the pre-pregnancy period.

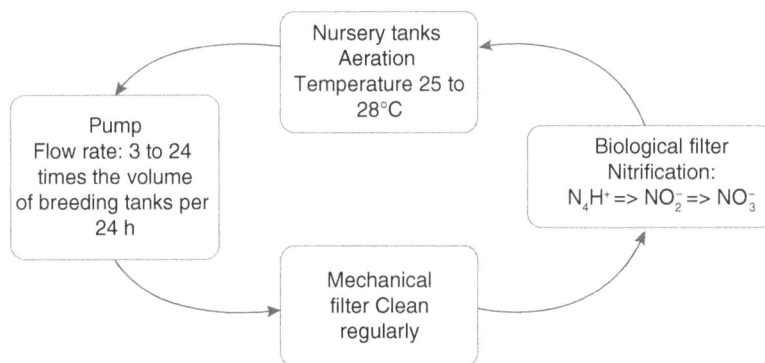

Figure 2.3. The main elements of the recirculating air system (RAS) in a nursery.

Table 2. Nursery dimensions for a farm with 1 ha of grow-out ponds.

Objectives of the grow-out pond
Pond rearing density: 3 to 3.5 shrimp/m².
Age of PLs introduced: 60 days (0.2 to 0.5 g)
Number of PLs60 (juveniles): 30,000 to 35,000

Nursery parameters:
Water temperature: 25 to 28°C
Expected survival rate over 2 months: 80%.
Number of PLs to be introduced into the nursery: 40,000 (survival around 32,000 PLs)
Density in nursery tanks on introduction: 2,000 PLs/m³
Presence of substrate: yes, 1 m² for 400 shrimp (including bottom and sides of tanks)
Bin dimensions:
Total volume of nursery tanks: 20 m³
Suggested number of rearing tanks: 2 tanks of 10 m³ each, or 4 tanks of 5 m³
Size of building:
A building or greenhouse of around 50 m² is required.

The size of the tanks can vary from a few m³ to 20 m³ or even 40 m³, depending on the capacity of the nursery. The recommended depth of the tanks is 1 m.

It is preferable to spread the LPs over several bins to limit the risk of loss in the event of an accident on one bin.

You will below further specific instructions concerning the use of PLs in nurseries.

2.5.1.3. Types of nursery tanks ⑥

Circular or sub-square tanks are recommended because they allow even water circulation. They should allow complete emptying through the bottom. Nursery tanks are ideal, in polyester or EPDM liner. For smaller budgets, 10 m³ leisure pools may be suitable, although they are not recommended because of the possible toxicity of the liners. However, the short duration of the nursery cycle limits the risk of toxicity, and some pools have been used successfully (Figure 2.4).

Figure 2.4. 10 m³ nursery tanks made from leisure pools and installed in a greenhouse.

2.5.1.4. Building

The breeding tanks are installed in a specific building or room. In temperate regions, this area must be insulated to limit heat loss and therefore the cost of heating the water. This can be a greenhouse, which has the advantage of providing heat on sunny days; however, it will need to be insulated either with double walls or with insulating plastic (bubble wrap). Bubble wrap can also be laid directly over the water in the tubs at night.

The building should be large enough to accommodate the rearing tanks and ancillary equipment, and to allow people to walk around and work in the area. Around 40% of the surface area of the building should be set aside for the tanks. For example, a greenhouse housing 40 m² of tanks would be 100 m². The building should preferably be lit by natural light. However, sunlight should not be allowed to shine directly onto the rearing tanks, in order to limit the greening of the water through the development of phytoplankton. This can be remedied by shading the greenhouses. Although phytoplankton can be useful in fish farming, particularly for recycling nitrates and providing nutritional supplements, it is not suitable for RAS systems because it clogs the filters.

2.5.1.5. Other equipment

To operate a recirculation system, the rearing tanks need to be equipped with a pump and a mechanical and biological filtration system. Within the nursery, each tank can be equipped individually with its own pump and filtration system. It is also possible to install a common filtration module shared by several tanks. Each system has its advantages and disadvantages.

Mechanical filtration is carried out using a sand filter or bag filter. The biofilter will have a volume representing 5 to 10% of the volume of the rearing tank(s). The biological filtration media can be static, ceramic or other porous material. The use of a fluidised bed with polyethylene (PE) or polypropylene (PP) biofiltration rings offers high efficiency and has the advantage of preventing clogging. Abundant aeration of the biofilter is then necessary.

The tanks should be aerated using a compressor and air diffusers. Heating is an important element, with the water temperature needing to be at least 25°C and optimal at 28°C. Heating is a major cost in nurseries, so cost-effective systems such as heat pumps are preferable.

Supports (substrates) for the shrimp will be provided to optimise the density of the PLs; these consist of nets or netting installed in the tanks ⑪.

Finally, the nursery must have analysis equipment available: thermometers, colorimetric kits for chemical analysis of the water (NH_3, NO_2, hardness), pH meter and oximeter.

2.5.2. Earthen grow-out ponds

Semi-extensive grow-out ponds are dug directly into the ground (Figure 2.6). The bottom of the pond is not covered with a synthetic membrane or concrete;

Figure 2.5. Essential equipment in a nursery tank. *From left to right*: resistance heater, membrane compressor, sand filter.

Figure 2.6. Shrimp grow-out ponds.

it is natural, made of earth. The soil must be sufficiently clayey to ensure that the ponds are watertight. As well as considerably reducing costs, natural soil provides the nutrients needed to start the food chain, with water/soil exchanges releasing trace elements. The soil provides a home for the many organisms in the pond that will help to create an ecosystem that is conducive to the development of shrimp.

2.5.2.1. Choice of site

The ideal site probably doesn't exist. It is possible to adapt pre-existing fish ponds, which will represent a substantial saving by limiting earthworks costs. When choosing a site, the following factors should be taken into account:

–   The climate must be conducive to the growth of *M. rosenbergii*, with water temperatures at or above 20°C for as long as possible, i.e. around 120 days continuously. These zones are located in southern Europe at low altitude. In mainland France, the south of the country is favourable, with altitudes of less than 300 or 400 metres. Areas with a continental climate and very hot summers, such as eastern France, may also be suitable.
–   Waterproofing the soil is essential. Before building any new pools, you need to make sure that the soil is suitable for retaining water. Sand or gravel should be avoided. Clay and silty-clay soils are best. Granulometry analyses based on soil samples taken at several points and at depth (1.50 m minimum) will ensure that a site is suitable.

- For the construction of new ponds, the land must be of a size suited to the production forecasts; a plot of land of less than 2 or 3 hectares is unlikely to be profitable. The land should be flat or slightly sloping to limit earthworks costs.
- Access to quality water, suitable for the species *M. rosenbergii*, is necessary. The annual volume needed to replace water lost through evaporation or infiltration must be calculated. The water source may be surface water, a river for example, or borehole water. Before any work is carried out, it will be necessary to check with the local authorities that it is possible to obtain authorisation to draw water.
- Access to electrical power will have to be taken into account. The site requires an electricity supply for the pumps and aerators, and possibly to heat the water in the nursery.
- Road access is required. If the project envisages direct marketing of the production, the visibility and accessibility of the site to the public must be taken into account, including parking facilities.
- Finally, proximity to the market can be an advantage, depending on the size and nature of the project (direct sales or sales to intermediaries). Proximity to a major urban centre is an advantage, as there will be plenty of customers and sales will be easier.

2.5.2.2. Basin development [7]

The grow-out ponds are made of earth. As we saw earlier, the nature of the soil is essential; it must be naturally impermeable. New ponds are built by earth-works; they are of the "dug and dyked" type. Part of the land is dug up and the surplus earth is used to create the dykes. Good compaction of the bottom and banks will determine the lifespan of the reservoirs.

The ideal shape for a freshwater shrimp pond is rectangular. The surface area can vary from 0.1 to 1 ha. A size of 0.2 to 0.5 ha is recommended, as it allows effective management and sufficient yield to justify emptying.

Pools are designed to drain efficiently. The bottom should have a slope of 0.3 to 1% and should be as smooth as possible. Even very slight depressions should be avoided, as they create puddles during emptying and give refuge to shrimps, which hampers harvesting efficiency.

The dikes have a gentle internal slope of 1:3 (1 m high by 3 m long) to limit erosion and facilitate access on foot. The external slope can be steeper: 1:2 or even 1:1. It is important to leave enough space for a vehicle to travel between the basins, so the recommended width of the top of the dykes is 3 m.

The height of the water must be between 0.7 and 1.5 m. The average is 1 m. So a pond could be 0.9 m deep at its highest point and 1.20 m deep at its lowest. This shallow depth is sufficient for shrimp; it also allows the water to warm up quickly on sunny days and allows the aquaculturist to walk around the pond. At this depth, you need to add about 30 cm of freeboard to the height of the dyke; this is the height between the water level and the top of the dyke.

Emptying is carried out at the lowest point of the basin, using a dedicated infrastructure. This infrastructure has two uses: to maintain the desired water level in the pond and to empty the pond at the required time. For large ponds (greater than 0.5 ha), this can be a monk. Two types of monk can be considered: internal or external (Figure 2.7). The monk consists of a concrete tower placed on a base (at the lowest point of the pond for an indoor monk). One of the walls of the tower is open and has grooves that allow two series of wooden planks to be slid through, sealed by clay placed between the planks.

For smaller pools, a tilting PVC stanchion is sufficient and less expensive (Figure 2.8). The water level is regulated by the central stanchion, which can be tilted to one side using an elbow to lower the water level. The addition of a sleeve, a PVC tube wider and higher than the central stanchion, allows the water to be drained from the bottom.

Whatever the structure used, a drainage pipe is placed at the bottom of the pool and crosses the dyke. It should have a sufficient slope to facilitate drainage and be fitted with one or more watertight collars to prevent water infiltration. Its diameter should allow the pool to be drained in less than 48 hours. In temperate zones, it can be urgent to drain a pool quickly at the end of the season

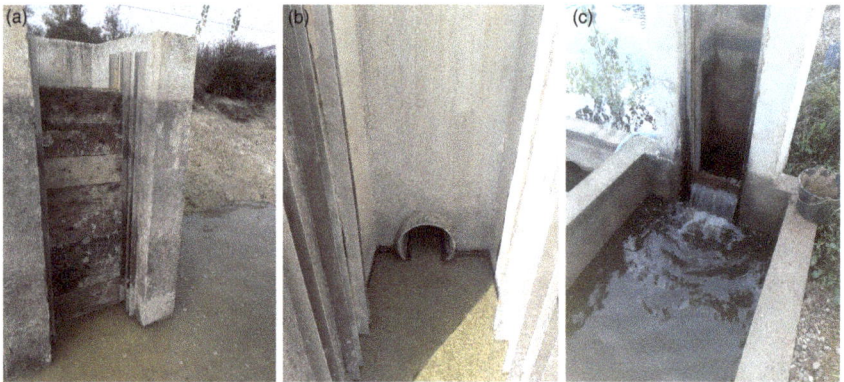

Figure 2.7. Inner monk with boards in an empty basin (2.7.a, *left*), inner monk without boards, with outlet pipe (2.7.b, *centre*) and outer monk integrated into a fishery (2.7.c, *right*).

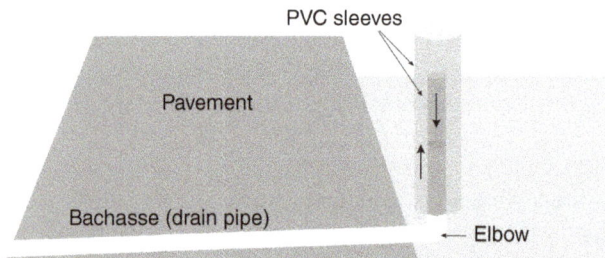

Figure 2.8. Principle of the PVC tilting stanchion.

if there is a sudden onset of cold weather. Rapid emptying will avoid losing a batch to cold mortality. The recommended diameters are 15 cm for a 1,000 m² pond, 20 cm for a 2,000 m² pond and 25 cm for a 5,000 m² pond.

2.5.2.3. Equipment

The grow-out ponds must be equipped to receive shrimp.

The first step is to plan the harvest. It is possible to harvest shrimp in the deeper zone located at the level of the drainage system (see Figure 2.9). However, in order to facilitate operations and preserve the shrimp, it is recommended that the site be equipped with one or more concrete fisheries located outside the ponds at the outlet of the drainage pipes. A concrete fishery can be shared between several ponds if it is strategically positioned. It will be between 5 and 20 m² long and can be compartmentalised (see, for example, Figure 2.10).

The installation of anti-bird nets is essential. It is required to obtain the status of closed aquaculture facility. The nets help to limit losses due to bird predation (cormorants in particular). The rectangular shape of the ponds makes them easy to install.

In order to optimise the growth of individual shrimp, the installation of substrates is of interest (Figure 2.11). These are nets laid vertically (possibly horizontally) inside the ponds. *M. rosenbergii* is a benthic species and does not tolerate high densities. The use of substrates increases the surface area available to shrimp by using the water column, thereby reducing the density of shrimp per unit area. The addition of substrates increases yields by 10 to 20% for the same number of shrimp reared. A 25-50% increase in surface area is recommended. For example, a 5,000 m² pond could be fitted with 1,250 to 2,500 m of 1 m vertical netting attached to stakes. Any type of support is acceptable: construction site nets are inexpensive, but they are made of non-recyclable plastic. Heavy shade cloth can be used; over time, it will be colonised by algae

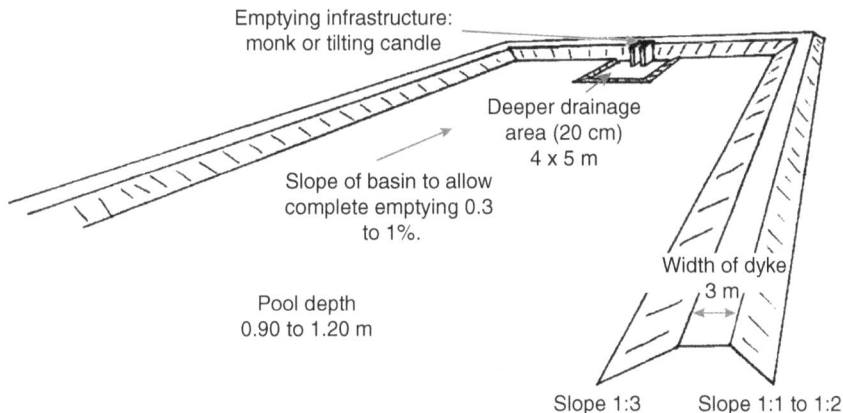

Figure 2.9. Characteristics of an earthen grow-out pond of 2,000 to 5,000 m².

Figure 2.10. Fisheries with two compartments (Gers, Gascogne Aquaculture).

Figure 2.11. Substrates made from construction site netting attached to cables stretched between acacia stakes.

and various micro-organisms (*fooling* phenomenon) and will contribute even more to the nutritional intake of the shrimps.

It is strongly recommended that ponds be equipped with an aeration system for emergencies. The level of dissolved oxygen required for good shrimp growth (> 3 mg/l) is usually available naturally in the low-density ponds typical of semi-extensive systems. However, this level can fall below the critical threshold of 1 mg/l, particularly at the end of the night when there is a *bloom of* phytoplankton or when it is very hot. An aerator can then be used to save the shrimp batch. Blade aerators are the most commonly used in shrimp farming. As well as oxygenating the water, they can also be used to stir it up and homogenise the water in the pond. In extreme cases, a more powerful aeration system such as a hydro-ejector may be required.

Finally, the standard equipment used in pond fish farming should be considered: landing nets for catching fish, pumps and small water analysis equipment. To capture samples of shrimp in open water in the ponds, for example for weight monitoring, you need to use large 1-metre landing nets, adapted to shrimp fishing, such as kissing nets or bichette nets.

2.5.2.4. Designing a closed-circuit site

The design of a closed aquaculture facility within the meaning of Regulation 708/2007 ❹ requires that there be no direct connection with natural waters. This means that water cannot be supplied to the ponds directly from a stream or natural body of water. Pumping the inlet water through a small-diameter screen (a few mm) will limit the possible entry of predators or undesirable fauna (crayfish, for example). Similarly, the basins cannot be emptied directly into natural water. The basins will have to be designed in such a way as to allow water to circulate between them, but not towards the outside environment. ❽ The aim is to avoid the risk of shrimp escaping into the natural environment during rearing, but especially when emptying the ponds.

Shrimp harvesting in a discontinuous system requires each pond to be emptied completely at the end of the grow-out season. It is therefore advisable to design the ponds at different heights so that they can be emptied by gravity from one to the other. The diagram in Figure 2.12 shows an installation with three 0.5 ha ponds each located at different heights. Pond 2 is the highest and can be emptied into either pond 1 or pond 3. Pond 3 is the lowest (pond 1 is at an intermediate height) and will collect the water drained from ponds 1 and 2. A fishery can therefore be built downstream of these basins. To be drained, basin 3 will need to have its water pumped upstream into 1 and 2. To prevent overflow in the event of excess water, the furthest downstream basin should be fitted with a gravel filtration system. If it is not possible to build basins of different heights, for example in the case of adapting existing fish farm basins, there will be no choice but to pump out each basin and reserve an empty or partially empty basin to receive excess water.

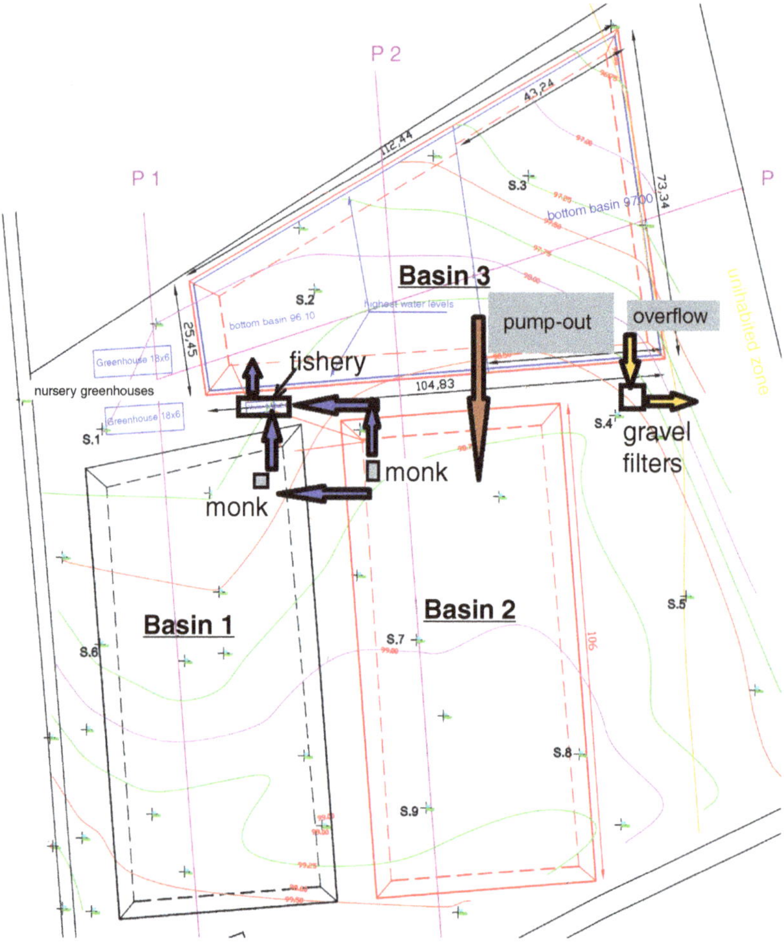

Figure 2.12. Diagram of a closed aquaculture facility with a drainage circuit.

An enclosed aquaculture facility will have to be built outside flood zones and will have to be fully fenced to limit the risk of theft or vandalism. Anti-bird netting will also be required.

Apart from the regulatory aspect, the closed facility has a number of advantages that are beneficial to *M. rosenbergii* breeding:

- it limits the risk of wild pests or predators being introduced into the breeding ponds;
- it reduces water consumption, thereby helping to preserve the natural aquatic environment;
- it limits the risk of chemical contamination (pesticides, etc.) and biological contamination (pathogenic micro-organisms) which can be fatal to farmed species;
- it aims to maintain a good health status for farmed species.

Limitations and other risks should also be taken into account.

In the event of heavy rainfall (rare in the summer in France), the design of the basins must avoid receiving run-off water (no connection with drainage ditches or streams). The basins, or the one furthest downstream, may overflow; prevention involves installing an overflow system fitted with a gravel filter.

Conversely, a long period of drought can lead to a shortage of water. Appropriate means of coping with this must be anticipated. Ideally, the site should be located close to a permanent watercourse from which it will be possible to pump water. Having the option of pumping from a borehole could be another way of dealing with the problem.

Notes

[1] *Recirculating Aquaculture System.*

[2] The term "juvenile" is used for PLs that are several weeks old when they are released into grow-out ponds.

Nursery Phase Management   3

The nursery is the intermediate stage between the hatchery and grow-out in outdoor ponds. In temperate zones, this phase is essential to increase the total production time, given the short summer period favourable to the growth of this species outdoors.

The nursery phase allows preliminary grow-out and increases the robustness of juveniles before they are moved outdoors, thereby limiting mortality due to stress and predation when they are placed in outdoor ponds. During the nursery phase, the *M. rosenbergii* population is able to support high densities, of the order of 1,500 to 2,000 individuals per m^3 (compared with the adult stage: 3 or 4 individuals/m^3) and to grow relatively evenly.

Management in the nursery phase is detailed in 4 data sheets: setting up post-larvae ⑨, daily management and water quality parameters ⑫, feeding ⑬ and monitoring and optimising growth ⑭. The chronology of the steps to be carried out in the nursery is shown in the frieze in Figure 3.1.

3.1. Placement of post-larvae in nursery ⑨

3.1.1. Objective

Controlled installation of post-larvae (PLs) in nursery tanks to ensure their survival and a good start to growth.

3.1.2. System or species specificity

M. rosenbergii is a species whose larval cycle takes place in brackish water. In the natural environment, post-larvae swim up rivers to continue their growth cycle in freshwater. The nursery phase in temperate zones takes place in freshwater over a period of around 2 months. As this species is sensitive to density, the

    DOI: 10.1079/9781836993162.0003

Empty tank	Nursery tank in water without prawns	Introduction of post-larvae in the basins	Nursery tank in water with post-larvae shrimps	Introduction the post-larvae to outdoor pools

January	February	March	April	May	June

Empty nursery	Preparing the bins	Setting up PLs	Growth and monitoring of PLs	Transfer of juveniles to outdoor pools
Corresponding texts	6	9	12 13 14	16

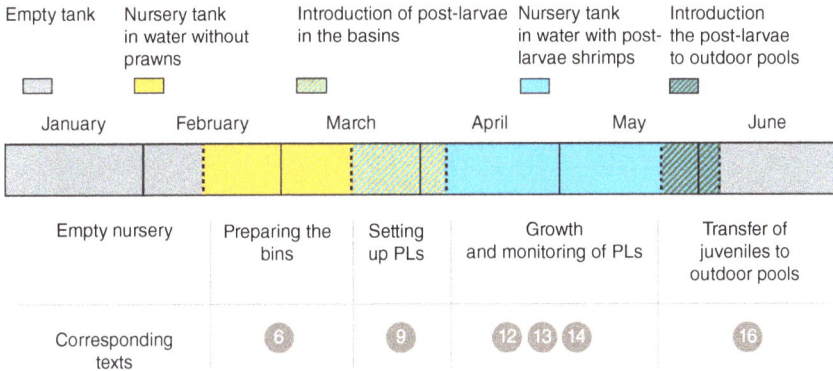

Figure 3.1. Chronology of activities in the nursery.

number of post-larvae per volume and unit area is an important criterion to consider when setting up a nursery.

3.1.3. Points to watch

– Post-larvae in good health on arrival: no or very few dead PLs; high activity, PLs moving in all directions.
– Post-larvae of uniform size (same age: within a few days).
– Pre-prepared tanks with activated biofilter and water at the required temperature (min. 25°C - max. 31°C).
– Volume of nursery tanks and substrate surface area to accommodate the number of PLs installed (see installation instructions).

3.1.4. Indicators of good practice

– Survival rate in the first 24 hours close to 100%.
– Optimum water temperature: 26 to 28°C.
– Low nitrite levels (< 0.2 mg/l); indicator of a biofilter in good condition.
– Substrates in place or ready to be installed.

3.1.5. Instructions for use

3.1.5.1. Acclimatisation to fresh water

The good survival of the post-larvae placed in the hatchery will depend to a large extent on prior and gradual acclimatisation to fresh water, carried out by the post-larvae supplier in the hatchery. It is suggested that the hatchery switch from a salinity of 12 to 0 PSU in at least 24 hours.

3.1.5.2. Acclimatisation on arrival in the nursery

The post-larvae are delivered in oxygen-inflated transport bags. The release of post-larvae on arrival should be done with care to avoid a sudden change in water parameters (temperature and pH). The temperature difference must not exceed 2°C between the water in the transport bags and that in the reception tanks. If there is a large difference (greater than 2°C), leave the transport bags floating on the surface of the receiving water for 30 minutes to 1 hour, so that the temperature of the water in the bags equalises. When opening the transport bags, mix for 15 to 30 minutes, gradually adding the water from the receiving ponds to the transport water.

3.1.5.3. Number of PLs to be introduced ⑩

For a nursery period of 60 days, the post-larvae will reach an average weight of 0.2 to 0.5 g/PL at D60. Placement at a density of 1,500 to 2,000 PLs/m3 is recommended (with no change in density over 60 days), and the substrate will correspond to a surface area of 400 PLs/m², including the surface area of the bottom and sides of the tank. The density can reach 5,000 PLs/m³ for the first 30 days. As a general rule, the biomass can reach 0.5 g/l and must not exceed 1 g/l; beyond this, growth and survival rates will be greatly reduced.

3.1.6. Recommendations and special precautions

–   Lightly salting the water (1 to 2 PSU) can limit stress when placing PLs in tanks.
–   Place a control batch in a basket to assess any mortality during the first few days (24 to 48 hours).
–   The substrate can be added after installation when the PLs grow and become more aggressive. ⑪
–   The substrate can be made up of netting (e.g. construction site netting) laid in the troughs. A cube-shaped substrate (1 x 1 x 1 m) is particularly effective. It is made from PVC pipe and plastic mesh or large-mesh shade cloth is attached every 5 to 10 cm (Figure 3.2).

3.2. Daily management and water quality parameters ⑫

3.2.1. Objective

Maintaining the physical and chemical parameters of the rearing water within the ranges suitable for *M. rosenbergii* in order to optimise growth and avoid excess mortality in the shrimp.

Figure 3.2. Nursery tank with two types of substrate: cube-shaped mesh substrate, left; site net substrate, right.

3.2.2. System or species specificity

M. rosenbergii is a tropical species; temperatures that are too cold (< 20°C) severely restrict shrimp growth and can cause death (< 13°C). Studies have shown that growth increases with temperature up to . The system recommended for nurseries in temperate zones is a recirculating circuit with a biofilter ("RAS"). If the biofilter malfunctions, excessive ammonia (NH_3) and nitrite (NO_2), which are toxic to shrimp, are produced. Concentrations of NH_3, NO_2 and NO_3 (nitrate) change over time as the biofilter is activated, with an ammonia peak in the first week, followed by a nitrite peak for 15 to 30 days.

3.2.3. Points to watch

- Poorly activated biofilter unable to degrade NH_3 and NO_2 effectively.
- Water too green due to excess phytoplankton.
- Overfeeding, visible as an accumulation of food at the bottom of the pond and responsible for the deterioration in water quality.
- Low outside temperatures (cold spells).
- Temperature variations during water changes.

3.2.4. Indicators of good practice

The physico-chemical parameters of the farm water are within the thresholds for the species (Table 3).

3.2.5. Instructions for use

3.2.5.1. Instructions for measuring water parameters

For each parameter of interest, these instructions concern the measurement tools, the recommended frequency and time of measurement, and the target values; they are grouped together in the following Table 4:

3.2.5.2. Instructions for corrective actions relating to parameters

- Temperature: if there is any deviation from the target temperatures, identify the cause and correct it. This could be a fault in the heating unit, an obstruction in the water flow, etc.
- Nitrite: if the nitrite concentration exceeds 1 mg/l, change the water regularly while waiting for the biofilter to stabilise, which may take more than 10 days. Water changes should not lead to a sudden change in pond water temperature of more than 3°C. If food has accumulated at the bottom of the pond, remove it by siphoning.

Table 3. Recommended, stressful and lethal levels of physico-chemical water parameters for rearing *Macrobrachium rosenbergii* (New, FAO, 2002).

Parameter	Optimum for the species	Stress level (S) or lethal level (L) for juveniles
Temperature (°C)	> 28 and < 31	< 19 (S)
		< 12 (L)
		> 35 (L)
pH (unit)	> 7.0 and < 8.5	> 9.5 (S)
Dissolved oxygen (ppm)	> 3 and < 7	< 2 (S) and < 1 (L)
Salinity (psu)	<10	
Turbidity (cm Secchi disc)	> 25 and < 40	
Ammonia NH_3 (ppm)	< 0.3	> 0.5 at pH = 9.5 (S)
		> 1.0 at pH = 9.0 (S)
		> 2.0 at pH = 8.5 (S)
Nitrite NO_2 (ppm)	< 2,0	> 3.0 (L)*
Nitrate NO_3 (ppm)	< 10	
Alkalinity (ppm)	20–60	
Total hardness (ppm $CaCO_3$)	30–150	

*Observation in Gascogne Aquaculture nursery.

Table 4. Measurement of physico-chemical water parameters.

Parameter	Measurement tools	Recommended measurement rate	Time of measurement	Target value
Temperature	Manual or automatic probe thermometer	2 times a day	Early morning and late afternoon	26 to 28 °C
Nitrite	Colorimetric test	1 time every 2 or 3 days*	If in doubt, at any time	< 0.3 mg/l
Dissolved oxygen	Electronic probe oximeter	When in doubt		> 5 mg/l
pH	Strip or probe pH meter	2 to 3 times a week	End of day	7 to 8.5
Hardness	Colorimetric test or strip	Bi-weekly or monthly		30 to 150 mg/l

*If a nitrite peak is identified (increase in nitrites above 1 mg/l), measure nitrites twice a day, morning and evening.

- **Dissolved oxygen:** a drop in oxygen is due to insufficient aeration. Check that the air compressor and diffusers are working properly, and take appropriate action.
- **pH**: change the water to the correct pH. Add bicarbonate of soda to raise the pH (if it is too low).
- **Hardness:** if hardness is too low, add a little lime or calcium carbonate to obtain a hardness of over 30 mg/l.

3.2.6. Recommendations and special precautions

- The bottom of the ponds should be regularly siphoned to remove food scraps and waste; cleanliness is essential to maintaining good water quality.
- A sudden increase in water turbidity due to a *bloom* (rapid, massive bloom) of phytoplankton can cause a rapid increase in nitrites at their death, which quickly becomes toxic for shrimps. Preventative water changes should therefore be made before the nitrite peak appears. Sufficient shading of ponds is the best way to prevent greening of the water.
- Keeping the water slightly salty for the first few weeks of the nursery helps to reduce the toxicity of nitrites for the shrimps.
- You can slow shrimp growth to keep them in the nursery longer by lowering the temperature to between 22 and 25°C.

3.3. Nursery feeding ⑬

3.3.1. Objective

Optimise the growth of post-larvae by controlling the quantity and quality of feed distributed.

3.3.2. Species or system specificity

The pre-growth phase of *Macrobrachium rosenbergii* in nurseries is relatively short, lasting around 2 months. It is necessary to optimise growth to obtain shrimps of sufficient size (0.2 to 0.5 g) when they are put into grow-out ponds.

Post-larvae are omnivorous, with a varied natural diet consisting of invertebrates and various plant debris. The ingestion, digestion and excretion cycle takes around 4 hours.

3.3.3. Points to watch

- Feed composition adapted to the species (see below).
- Good quality food, within use-by dates.
- Avoid under- or over-feeding.

3.3.4. Indicators of good practice

– PL growth curve in line with the reference curve (Figure 3.3).
– Clean bottom of bins, with no food residue.
– No nitrite peak.

3.3.5. Instructions for use

3.3.5.1. Composition of the feed

Inert feed with a low moisture content is recommended. It is distributed in the form of 0.4 to 0.7 mm crumbs for the first 6 weeks, then 1 mm pellets thereafter. Table 5 shows the composition of the feed recommended for PLs in nurseries.

3.3.5.2. Feed distribution

3.3.5.2.1. NUMBER OF DAILY DISTRIBUTIONS: Feed should be distributed at least twice a day. Continuous distribution would be optimal to satisfy nutritional requirements and optimise growth, but also to limit cannibalism. It is possible to divide up the ration using dispensers positioned above areas of high bubbling.

3.3.5.2.2. DISTRIBUTION TECHNIQUE: Calculate the quantity of feed to distribute per tank using the reference feeding table (Table 6), based on the estimated number of individuals and their average weight. As shrimp do not move very much to look for food, it is necessary to distribute the feed evenly over the entire surface of the tank. Distributing it in areas where there is a lot of bubbling will help to disperse it throughout the water mass. The food will then be better used, as it will be ingested more quickly and will not have time to leach out.

3.3.5.3. Feeding table

The following feeding table (Table 6) is based on an average water temperature of 25°C. If the average temperature of the tanks is 28°C, the feed dose should be increased by around 30%.

Table 5. Recommended composition of a feed for post-larvae in nurseries.

Component	Recommended rate
Protein (in %)	> 35 %
Fat (in %)	4 to 8 %
Digestible protein/digestible energy (in mg protein/kcal)	> 17
Fibre/cellulose (%)	0 to 30 %
Vitamin C (in mg/kg)	> 135

Table 6. Nursery feeding table for *Macrobrachium rosenbergii* at 25°C (data from Gascogne Aquaculture).

Age of post-larvae (in days)	1	8	15	22	29	36	43	50	57	64	71
Weight of post-larvae (in g)		0.006	0.01	0.016	0.025	0.035	0.05	0.075	0.1	0.15	0.25	0.35
Feed (g/d/1,000 PLs)	1	1.3	1.5	2	2.6	3.5	4.8	6	8	10	14
Feeding rate (%)	16.7	13	9.4	8	7.4	7	6.4	6	5.3	4.8	4

3.3.6. Recommendations and special precautions

–   Regularly check the bottom of the rearing tanks for food remains. If this is the case, the quantity of feed distributed is excessive, or the PLs are experiencing discomfort (stress linked to poor water quality, illness, etc.). Reduce the amount of feed and correct the cause of discomfort, if necessary.
–   Daily siphoning allows the ration to be adjusted by estimating the amount of leftovers. Bubbling creates convection zones; areas where leftovers accumulate are relatively stable and can be cleaned even blindly.
–   It is not essential to use a specific feed for prawns. Trout nursery feeds give very good results (size AL1 and AL2).
–   The use of fish or shellfish in the preparation of a household ration is possible and has been described by several authors. However, this presents a significant health risk of introducing disease into the farm and/or degrading water quality. For these reasons, a home-made ration is not recommended.

3.4. Monitoring and optimising the growth of PLs in nurseries ⑭

3.4.1. Objective

Maintain optimum weight growth and survival rate of PLs during the nursery phase by applying regular weight monitoring and appropriate management procedures.

The recommended nursery period in temperate zones is 2 months (60 days). Provided the water temperature is maintained at or above 25°C and the density is reasonable, the average weight of juveniles reaches 0.2 to 0.5 g in 60 days.

3.4.2. Species or system specificity

M. rosenbergii does not tolerate high densities. Heterogeneous individual growth is a particular feature of this species: some individuals grow faster and inhibit the growth of dominated individuals. In nurseries, this characteristic is less marked than in grow-out, but must be taken into account. The number of post-larvae per volume of water and per unit area (densities) is one of the criteria to be controlled in nurseries. Growth decreases and cannibalism increases when the biomass density of PLs exceeds 0.5 g/l.

3.4.3. Points to watch

– Tank volume and substrate surface area adapted to the number and average weight of PLs throughout the nursery phase.
– Monitor and prevent cannibalism (over- or under-feeding).
– Controlling water parameters, particularly temperature.
– Monitor biomass: growth decreases and cannibalism increases when the biomass density of PLs exceeds 0.5 g/l. It should never exceed 1 g/l.

3.4.4. Indicators of good practice

– High survival rate at the end of the nursery period: 80 to 90%.
– Substrates in place allowing a maximum density of 400 shrimp/m^2 at 60 days.
– Water parameters continuously within optimum values, particularly temperature ⑫.
– Good food intake observed. However, avoid overfeeding.
– Uniform PL size, as far as possible.
– Growth curve similar to the reference curve (Table 7 and Figure 3.3).

3.4.5. Instructions for use

– **Regular measurement of the average weight of each batch in the nursery**. Weighing is carried out once a week if possible, or once a fortnight if this is not possible. The data is recorded and compared with the reference growth curve (Table 5 and Figure 3.3).
– **Weighing procedure**: samples of around twenty PLs per batch are weighed using a dip net. A precision balance (0.01 grams) was used. The prawns

Table 7. Growth of post-larvae in nursery at 25°C, with an initial density of 2,000 PLs/m^3 (data from Gascogne Aquaculture, 2021).

Age (in days)	1	8	15	22	29	36	43	50	57	64	71
Weight (in g)	0.006	0.01	0.016	0.025	0.035	0.05	0.075	0.10	0.15	0.25	0.35

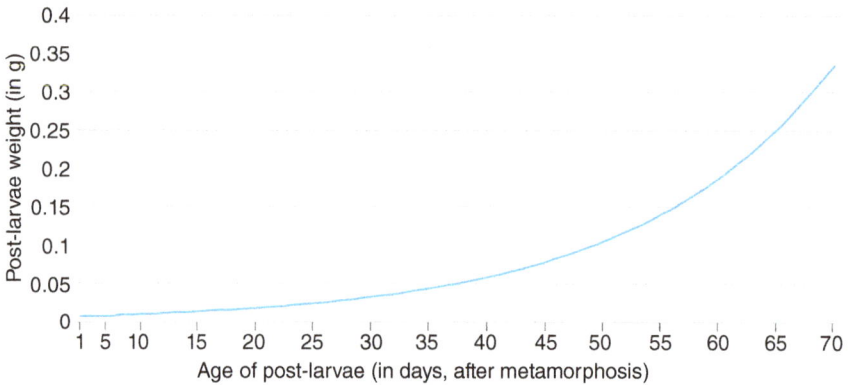

Figure 3.3. Reference growth curve for post-larvae in nurseries at 25°C.

are placed dry in the net and can be weighed dry or in a small volume of water after taring the balance.
– **Sorting PLs** reduces the effect of heterogeneous individual growth by separating fast-growing individuals. The yield (growth and survival) is then improved for all batches. This sorting is carried out using a sorting grid (Figure 3.4) in the second half of the nursery period (D30 to D50) and/or during the transfer (D60) to the outdoor pond.
– **The addition of substrate** is necessary if the initial surface area is insufficient, the aim being to achieve a density of 400 shrimp/m^2 at D60 ⑪.

3.4.6. Recommendations for special precautions

– Sorting grids: 2 to 6 mm apart depending on the size of the PLs. It is recommended to sort with a 4 mm grid at D60 to separate a batch of PLs with an average weight of around 0.4 g.
– Sorting one lot into two new lots requires the necessary infrastructure to be provided to accommodate them (doubling the number of bins).
– The higher the average weight of juveniles at the end of the nursery, the better the yield in outdoor grow-out ponds. The success of the nursery phase will have a direct impact on the overall economic performance of the business.
– The following factors help to improve growth in the nursery:
   • temperature: faster growth at 28°C than at 25°C;
   • density: the lower the batch density, the better the growth and survival rate;
   • sorting PLs: homogeneous batches grow better;
   • the duration of pre-pregnancy: extending the nursery phase to 3 months or more can produce large juveniles (over 1g); however, the density needs to be lowered considerably, which requires large volumes and increases the cost.

Table 8 and Figure 3.5 show the yields (survival and average weight) obtained during experiments at the ONIRIS-INRAE aquaculture station in 2021 in 200-litre tanks at different densities and over a period of up to 104 days.

– By stocking batches of different weights in dedicated ponds for outdoor grow-out, the final harvest can be staggered. Ponds stocked with large juveniles will reach market size earlier and will be harvested first. This will extend the shrimp sales period.

Figure 3.4. Sorting post-larvae using a grid.

Table 8. Survival rate of PLs in nurseries as a function of density at D74 and D104 at a temperature of 26°C (ONIRIS/INRAE experiment in 2021, Max Guézou).

	Density (PLs/m²)		Age of PLs (in days)	
Bins (200 l each)	per unit area (PLs/m²)	per unit volume (PLs/m³)	74 d.	104 d.
CLEAR 50	50	500	84 %	74 %
RAS 100	100	1 000	78 %	67 %
RAS 200	200	2 000	71 %	58 %

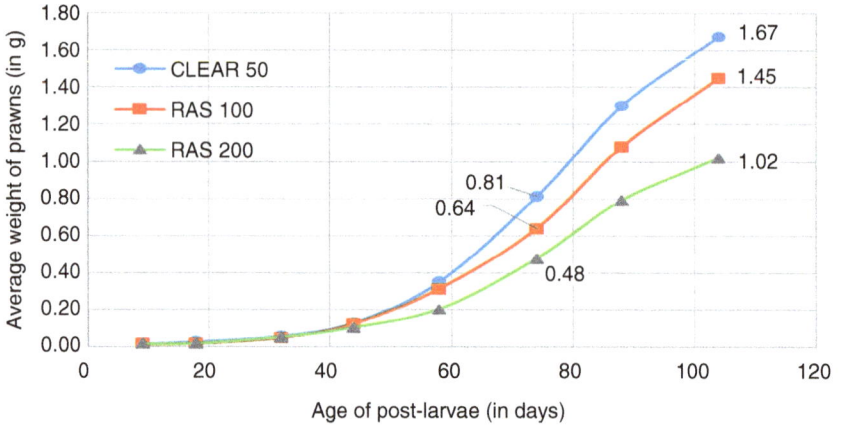

Figure 3.5. Evolution of the average weight of PLs in nursery tanks at different densities, t = 26°C (Oniris/INRAE experiment in 2021).

Management During the Grow-out Phase in Ponds

4

Grow-out in outdoor ponds is the final stage in the production cycle, ending with the harvesting of shrimp for marketing. In temperate zones, this phase lasts a short time, around 4 months, dictated by the favourable temperature for growth in summer.

Management during the grow-out phase is detailed in 7 technical sheets: tank preparation ⑮, introduction of juveniles ⑯, daily water management ⑰, phytoplankton management ⑱, feeding and fertilisation ⑲, growth monitoring ⑳, and predator prevention ㉑. The chronology of activities during the grow-out phase is represented by the frieze in Figure 4.1.

4.1. Preparing the basins ⑮

4.1.1. Objective

To ensure optimum conditions for pond productivity and water quality before juvenile fish are placed in the pond.

4.1.2. Species or system specificity

In a semi-extensive system, shrimp feed on the trophic chain that is naturally established in the pond (plankton, invertebrates, etc.). Maintaining a good level of mineral resources is essential to the proper development of the food chain and therefore of the shrimp.

4.1.3. Points to watch

- Turbidity;
- Water quality parameters: pH, calcium concentration (alkalinity and hardness), temperature.
- Concentration of mineral elements in water (N, P).

DOI: 10.1079/9781836993162.0004

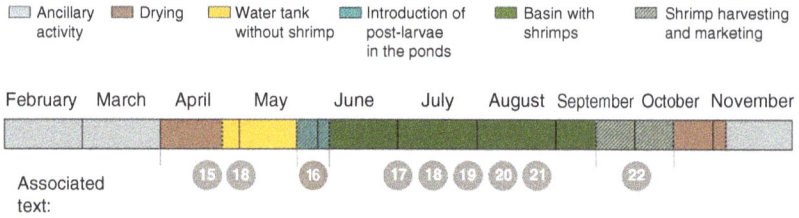

Figure 4.1. Chronology of magnification activities.

- Presence of macrophyte algae in the basins.
- Presence of predators or undesirable species that may carry pathogens (crayfish).

4.1.4. Indicators of good practice

- On the days preceding the introduction of the juveniles, water quality parameters adapted to the species *M. rosenbergii* ⑰, in particular turbidity, pH, alkalinity, hardness and temperature.
- Green-brown colour of the water.
- No mortality during the introduction of juveniles.

4.1.5. Instructions for use

4.1.5.1. Drying up

- The water should be drained at least once a year, preferably in April before the ponds are filled and the juveniles introduced, as this will eliminate predators such as insects (dragonfly larvae, dytics, etc.).
- It should last at least 2 or 3 weeks.
- In the event of heavy rainfall, drying out can be difficult and ineffective.
- It may be beneficial to work the soil mechanically (ploughing) during the dry period to increase the oxidation of reduced matter left over from previous rearing.

4.1.5.2. Filling the ponds

- Fill the ponds with water 4 weeks before the juveniles are due to be introduced.
- Take care to avoid introducing predators such as fish or insect larvae by using a grid ㉑.

–　Take care to avoid introducing crayfish or crayfish larvae, which can be present in natural waters from April onwards.
–　Ensure the quality of the water introduced and the absence of any pollution, depending on its origin.

4.1.5.3. Calcium amendment

–　If the pH of the water is acidic (< 7), add calcium in the form of lime or calcium carbonate. Spread between 500 and 1,000 kg/ha of quicklime on the pond soil (Figure 4.2).
–　If the land is not dry, apply up to 200 kg/ha of lime (preferably slaked) in several passes.
–　Be careful, these products spread in water can cause flocculation of phytoplankton and a significant drop in dissolved oxygen.

4.1.5.4. Organic fertilisation

–　Organic fertilisation should be carried out as soon as the pond is impounded. This will stimulate the development of phytoplankton, the basis of the pond's food chain.
–　The fertiliser used can be animal manure: preferably dehydrated cattle manure or poultry droppings. Manures are difficult to handle, and you need to use 2 to 4 tonnes per hectare of cattle manure (2 times less for droppings). Check their origin, as they may be contaminated with chemicals (anti-parasitics, antibiotics, etc.) or undesirable germs.
–　Rather than manure, organic fertilisation is best carried out using the feed that will subsequently be distributed to the shrimps, oilcake, alfalfa plugs or agricultural by-products [19];
–　The following protocol is suggested: start fertilising 4 weeks before introducing the juveniles, with a one-off application of 200 kg of feed per hectare. 4 days later, add 15 to 20 kg/ha of feed daily until adequate turbidity is achieved (less than 50 cm on the Secchi disc).
–　This protocol should be adapted according to the natural fertility of the soil in the basin and its history.

4.1.5.5. Mineral fertilisation

–　Mineral fertilisation should be avoided in the first instance, and organic fertilisation should be preferred to boost pond productivity.
–　In the event of insufficient phytoplankton development, mineral fertilisers enable rapid action, acting more quickly than organic fertilisers.
–　Mineral fertilisers can also rectify imbalances in N and P concentrations, giving an N/P ratio of between 4 and 10, with a dissolved PO_4 of between 0.2 and 0.5 mg/L.

- Nitrogen can be applied in the form of urea or ammonium nitrate (33.5), in the order of 20 to 30 kg/ha per application, and phosphorus in the form of ammonium polyphosphate (48.00), in the order of 15 to 20 kg/ha per application.
- Favour liquid forms, which are more effective and better adapted.
- Solid fertilisers can be diffused by attaching a permeable bag filled with the fertiliser to the front of a working blade aerator. The current generated will dissolve and disperse the fertiliser.

4.1.6. Recommendations and special precautions

- Draining the water has many benefits. It allows the organic matter accumulated at the bottom of the pond to be mineralised. It eliminates aquatic life, including potential predators and pathogens.
- A calcium amendment is not compulsory, but it does help to increase the pH, calcium content and hardness of the water.
- Liming can also be used to disinfect ponds: in this case, the protocol is to spread 1,500 kg/ha of quicklime on damp soil and leave it to work for at least 1 week before re-watering the pond.

Figure 4.2. Spreading lime on the floor of a dry pond.

– A lack of nitrogen can lead to the development of cyanobacteria to the detriment of phytoplankton. Cyanobacteria can cause an imbalance that is detrimental to the health of the pond.

4.2. Introduction of juveniles ⑯

4.2.1. Objective

Optimise the chances of juvenile fish surviving and starting to grow when they are placed in ponds.

4.2.2. Species or system specificity

M. rosenbergii is a tropical species that lives in warm waters. Below 19°C or 20°C, growth is greatly slowed and the vulnerability of the shrimp increases. In temperate zones, the temperature is below 19°C during winter and early spring. *Macrobrachium rosenbergii* cannot be grown outdoors all year round.

4.2.3. Points to watch

– Avoid thermal shock during transfer.
– Check the weather forecast before releasing the PLs outdoors; it should allow the water temperature to be suitable (>19°C) in the days and weeks that follow.
– Pond pH: avoid excessive differences with the nursery.

4.2.4. Indicators of good practice

– Less than 3°C difference in water temperature during transfers.
– No mortality due to transfer handling.
– No mortality in the outdoor ponds in the weeks following transfer.
– Temperatures, pH and dissolved oxygen levels within the tolerance ranges of shrimp ⑰, during and after transfer.
– Average weight of juveniles > 0.2 g.
– Uniform size of shrimp batches.

4.2.5. Instructions for use

– Seed the basins at a density of 2.5 to 3.5 PLs/m².
– The prawns are transferred to the temperate zone (south of France) at the end of May or beginning of June.

- Wait for the right conditions: weather forecast allowing water temperature above 19-20°C in outdoor ponds in the weeks following transfer, turbid water (Secchi disc between 25 and 50 cm), pH and oxygen within the tolerance range of shrimp ⑰.
- If the water parameters of the outdoor ponds and nursery tanks (T, pH) are too different, add water from the grow-out ponds to the nursery tanks before the transfer (> 1 hour) to adapt the shrimps.
- To catch PLs in nurseries: carefully remove the substrates, lower the water level and fish with a landing net.
- Weigh the prawns at the beginning, middle and end of the catch, as average weights vary (the largest remain at the end), and calculate an average.
- Sorting the PLs using a sorting grid at the time of transfer makes it possible to homogenise the batches placed in basins ⑭.
- Introduce the PLs at different points in the ponds (Figure 4.3); this allows the shrimp to colonise the whole pond more quickly.

4.2.6. Recommendations and special precautions

- Keep a sentinel sample of PLs (between 10 and 20 individuals) in a basket for a few days or weeks after transfer, in order to estimate any losses suffered by the batch.

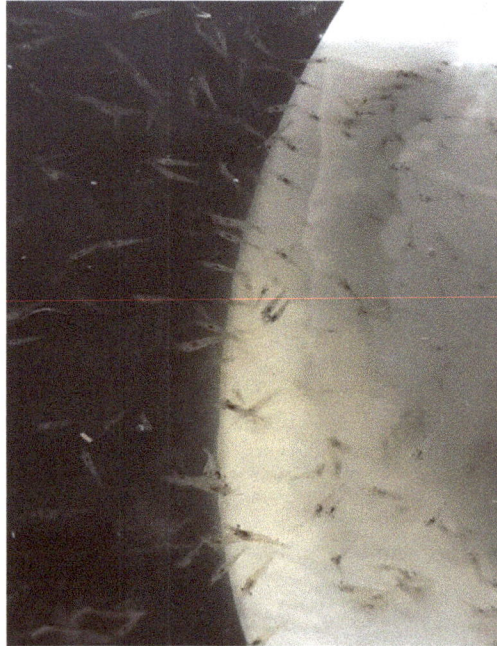

Figure 4.3. Releasing juveniles into a pond.

- The ideal weight of PLs when transferred to outdoor ponds is between 0.2 and 0.5 g. The larger the shrimp, the less sensitive they will be to stress or potential predators ㉑ and the better the final yields.

If sorting or transfer bins are used:

- Monitor the oxygen concentration in these ponds (above 3 mg/l)
- The high density in the tanks puts stress on the shrimp, so handling must be quick and gentle to avoid mortality.

4.3. Daily management and water quality parameters ⑰

4.3.1. Objective

Maintain the physico-chemical parameters of the water at levels suitable for *M. rosenbergii* in order to optimise shrimp growth and avoid mortality.

4.3.2. Species or system specificity

Water quality parameters, particularly temperature, have an impact on the growth of *M. rosenbergii*. Below , growth is slowed. Studies carried out in the United States have shown that growth is enhanced at temperatures below the optimum (28°C), but above 20°C, because this delays the expression of sexual and reproductive characteristics (egg-laying in females, aggressiveness in males) in favour of growth.

4.3.3. Points to watch

- Sudden drop in outside temperature.
- Sudden drop in dissolved oxygen levels at the end of the night, before photosynthesis resumes.
- Decline in shrimp growth.
- Heavy rain.
- Contamination by pesticide-laden water.
- Massive death of phytoplankton.
- Presence of cyanobacteria.

4.3.4. Indicators of good practice

Measurements of the physico-chemical parameters of the water, in particular temperature, must fall within the comfort ranges of the species as detailed in Table 2. The temperature in the temperate zone, the most important parameter, should be above 19-20°C for the duration of the grow-out.

Other indicators:

– minimum water height in the basin of 70 cm at the lowest point;
– shrimp growth curve similar to the reference curve (Figure 4.5);
– no run-off into basins from agricultural surfaces treated with pesticides.

4.3.5. Installation instructions:

4.3.5.1. Instructions for water parameters

Measurements of the physico-chemical parameters of the water are carried out according to the instructions detailed in Table 9.

Temperature, dissolved oxygen and pH can vary depending on the location in the water column. Measurements should be taken at depth, corresponding to the shrimp's preferred living zone.

4.3.5.2. Instructions for corrective action on parameters

– **Temperatures.** There are few techniques for heating the water in ponds. At the end of the season, shrimp should be harvested quickly if temperatures drop significantly. If the water is cool at the bottom and the days are sunny, use an aerator during the day to warm the water at the bottom, mixing it with the warmer surface water.
– **Dissolved oxygen.** The aim is to maintain a dissolved oxygen concentration above 3 mg/l at depth. Phytoplankton naturally produces oxygen during the day and consumes it at night; its management is developed below ⑱.

Table 9. Instructions for measuring the main physico-chemical parameters of water in grow-out ponds.

Parameters	Measurement tools	Measurement frequency	Time of measurement	Target value
Temperature	Hand-held, probe or recording thermometer	Up to 2 a day; important during periods of risk	Early morning and late afternoon	20 to 30°C
Dissolved oxygen	Electronic oximeter	Daily (depending on temperature and turbidity conditions)	Early morning	Between 3 and 7 mg/l
pH	Colorimetric test (ex pH strip) or pH meter	2 to 3 times a week	End of day	6.5 to 9
Hardness and alkalinity	Colorimetric test	Bi-weekly or monthly		20 to 150 mg/l
Turbidity	Secchi disc	Daily	Lunchtime	25 to 50 cm
Water height	Meter or permanent system installed in the pool	Weekly		Minimum 70 cm, target 1 to 1.2 m

The amount of dissolved oxygen is improved by the use of aerators. Blade aerators stir the water at the surface and are effective at homogenising pond water; they are recommended for semi-extensive shrimp aquaculture. Other powerful models, such as hydro-ejectors, can be used in emergencies, propelling fine air bubbles through a propeller using the Venturi effect. They should be used during the night (especially in the second half of the night) to compensate for the consumption of oxygen during the night by phytoplankton, particularly in the event of a phytoplankton *bloom*.

- **pH**. There are two main ways of controlling pH. By controlling phytoplankton ⑱ and by amendments ⑮.
- **Hardness**. Water hardness can be increased by spreading lime (preferably slaked) up to 200 kg/ha in several passes. It is also possible to amend the soil during the dry season to increase water hardness for the following season ⑮.
- **Turbidity**. Water turbidity is controlled by acting on phytoplankton ⑱.

For ponds in closed systems, it is important to allow for the possibility of adding water in the event of heavy evaporation of the pond water during the hot season. The water supplied, with the appropriate physico-chemical character-istics, can come from a river, a nearby pond or a borehole, and will be filtered through a grid to prevent predators from entering.

4.3.6. Recommendations, special precautions

- Measurements can be automated by installing automatic measuring devices. It is advisable to check the accuracy of these measurements regularly and manually.
- The solubility of oxygen depends on temperature. The higher the temperature, the less soluble the oxygen. Don't hesitate to ventilate in very hot weather.
- Good hardness and alkalinity limit variations in pH and are necessary to ensure the solidification of the exoskeleton following shrimp moulting. Calcium carbonate has a buffering effect on pH, which is very useful in shrimp farming.
- Turbidity in ponds can be phytoplanktonic or mineral (suspended sediment) in origin. Low turbidity, above 50 cm, indicates a lack of phytoplankton, which is favourable to the development of macrophyte algae. If the turbidity is too high, with a Secchi of less than 25 cm, there may be an excess of phytoplankton, which will consume oxygen during the night, causing its concentration to fall below a critical threshold.

4.4. Phytoplankton management ⑱

4.4.1. Objective

Control the development of phytoplankton to ensure optimum feeding and growth of shrimp.

4.4.2. Species or system specificity

In a semi-extensive system, shrimp are not fed directly with food, but feed on the natural productivity of the environment. Shrimp are omnivores, feeding mainly on zooplankton, algae and invertebrates. Phytoplankton is the first link in the food chain and depends on the nutrients provided by fertilisation for its development. Optimising shrimp growth therefore requires good phyto-plankton management.

4.4.3. Points to watch

- Presence of macrophyte algae.
- Concentration of mineral elements in water (N, P).
- Significant drop in dissolved oxygen overnight.
- Plankton *bloom*: excess plankton (Secchi < 20 cm).
- pH too high in the evening (> 9).
- Presence of cyanobacteria.

4.4.4. Indicators of good practice

- Turbidity measured by the Secchi disc between 25 and 50 cm.
- Uniform green-brown colour of the water.

4.4.5. Instructions for use

Phytoplankton are established by proper preparation of the pond ⑮ and adequate fertilisation ⑲.

4.4.5.1. Turbidity measurement:

Phytoplankton abundance is measured daily using a Secchi disc. If turbidity levels are too high or too low, action needs to be taken.

4.4.5.2. Establishment of phytoplankton

- Phytoplankton development begins as soon as the ponds are impounded, with organic (or mineral) fertilisation before the introduction of postlarvae ⑮. The elements nitrogen (N) and phosphorus (P) must be present.
- To obtain the optimum amount of phytoplankton, you need a dissolved N/P ratio of between 4 and 10, and a dissolved PO_4 of between 0.2 and 0.5 mg/l.
- Drying out allows the organic matter accumulated at the bottom of the pond to be mineralised, making mineral elements (N and P) available to phytoplankton.

4.4.5.3. Regulation of phytoplankton abundance

- If turbidity is too low (for example, a Secchi disc greater than 50 cm), phytoplankton development should be encouraged by increasing the amount of organic fertiliser (feed) distributed on a daily basis.
- If there are no results, mineral fertiliser ⑮ can be added on a rational basis. This has a rapid effect, as the nutrients (minerals) needed by the phytoplankton are almost instantly available.
- If turbidity becomes too high, when the Secchi disc measurement is less than 25 cm, daily feeding should be reduced or even temporarily stopped (if Secchi < 20 cm), until turbidity returns to a lower level.
- If the infrastructure allows, some of the phytoplankton can be eliminated by partially renewing the water in the pond. It is the surface water, which is highly charged with phytoplankton because it is in direct contact with light, that must be evacuated if the system allows (in the case of a monk or tilting candle). The evacuated water will be replaced by clearer water (river, borehole, pond).
- In the event of a major *bloom*, this water renewal can be carried out quickly by *flushing* in the afternoon, followed by replenishing the water during the night to limit the reactivation of phytoplankton.

4.4.6. Recommendations and special precautions

- Depending on variations in the environment (temperature, light, inputs, etc.), the quantity of phytoplankton can fluctuate and have an impact on other water quality parameters such as pH and dissolved oxygen.
- The lack of phytoplankton allows macrophyte algae to grow. These algae consume nutrients, preventing the phytoplankton from developing properly and hindering shrimp harvesting.
- The overabundance of phytoplankton means that there is a risk of a significant drop in dissolved oxygen during the night. Phytoplankton do not photosynthesise at night and consume oxygen, which can cause their concentration to fall below a critical threshold.

4.5. Feeding and fertilisation ⑲

4.5.1. Objective

Cover the nutritional needs of shrimp for optimum growth by distributing a feed that activates the natural productivity of the ponds.

4.5.2. Species or system specificity

M. rosenbergii has an omnivorous diet. In a semi-extensive system, the shrimp feed mainly on organisms produced naturally in the ponds (zooplankton, invertebrates, algae). The food distributed acts as a fertiliser and is not (or only to a limited extent) consumed directly by the shrimp. Fertilisers provide nutrients that are essential to the biological cycle and productivity of the pond.

4.5.3. Points to watch

- Protein content of feed/fertiliser.
- Chemical and biological contaminants.
- Significant drop in dissolved oxygen overnight.
- Plankton *bloom*: excess plankton (Secchi < 20 cm).
- pH too high in the evening (> 9).

4.5.4. Indicators of good practice

- Secchi disc measurement between 25 and 50 cm;
- Shrimp growth curve similar to the reference curve (Figure 4.5).

4.5.5. Instructions for use

4.5.5.1. Choice of feed

- **Agricultural by-products:** these are protein-oil cakes and other raw plant by-products usually intended for livestock feed. To facilitate feeding, they must be in pellet form. Alone or mixed, the average protein content should be 17-20% in low-input semi-extensive systems (New *et al.*, 2010). In the United States, soya meal, maize gluten and wheat remould are used. In the Gers, sunflower meal (Figure 4.4 and Table 10) has been used successfully, with or without alfalfa plugs.
- **Organic fertilisers:** organic manures of animal origin such as cattle manure or poultry droppings can be used in semi-extensive systems. Dehydrated products are available on the market and can be easily handled. These inputs are best used for preliminary fertilisation of ponds ⑮.
- **Manufactured feeds:** industrially manufactured feeds are optimally formulated to meet the nutritional requirements of each species. They are of limited interest in semi-extensive farming because of their high cost and questionable environmental impact (presence of fish meal). They would be of interest in farms with a density of more than 4 shrimp/m^2 (not promoted

in this book) and must then contain a minimum of 32% protein for *M. rosenbergii*. A feed for omnivorous fish such as carp, tilapia or catfish can then be used (Table 10). Feeds for penaeid shrimps are too rich in protein and will be underused.

– **Aquaculture feeds without fishmeal:** feeds based on cereals and other plants would provide more starch than highly fibrous cakes. Less expensive and more sustainable than industrial aquaculture feeds, they can be manufactured locally. Trials need to be carried out to assess their effectiveness.

4.5.5.2. Daily feeding

– The grow-out period lasts around 4 months. In the semi-extensive system, the total quantity of feed over this period is calculated from the maximum forecast harvest weight (1,000 kg/ha), multiplied by 2.5. Daily intakes are distributed according to the grow-out period, increasing progressively with the weeks, as shown in Table 11.

– Inflows need to be adapted to the turbidity of the basin. If turbidity is too low, you need to increase inputs, and vice versa.

– Shrimp have a marked territorial behaviour, so they are distributed throughout the pond and move very little to feed. The feed must be distributed so that it covers the entire surface of the pond.

– The feed is dispersed from the pond dikes using a feed scoop (Figure 4.4).

Table 10. Composition of different feeds used in rearing *M. rosenbergii*.

Components (% of total)	Organic sunflower meal, Gers origin	Mixture of alfalfa cap (70%) and sunflower meal (30%)	Industrial food (carp, tilapia, catfish)
Proteins	24 to 26 %	19.5 %	32 to 35 %
Lipids	16 %	2 %	9 %
Starch	Low (< 5%)	Low	19 to 25 %
Cellulose	18 to 22 %	21.5 %	4 %
Mineral matter	4.5 %	10.5 %	7 to 11 %

Table 11. Fertilisation table for ponds with a forecast production of 1 t/ha.

Weeks	of quantity over the period	Total quantity in kg/ha over the period	Quantity (in kg/day/ha)
1 to 5	20 %	525 kg	10 to 20 kg/d
6 to 10	25 %	612 kg	15 to 20 kg/d
11 to 15	30 %	787 kg	20 to 25 kg/d
16 to 18	25 %	630 kg	30 kg/d
Total	100 %	2,554 kg	

Figure 4.4. Sunflower meal pellets thrown with a feed shovel.

4.5.6. Special recommendations and precautions

–   The natural fertility of the ponds comes into play independently of the feed
    supplied; this fertility will depend in particular on the nature of the soil
    and the preparation work carried out by the grower [15].

–   It is possible to feed every other day, doubling the quantities of each feed.
    Yields do not seem to be affected.

–   At low densities (< 3.5 shrimp/m^2), a comparable yield was observed
    with a 20% protein feed based on agricultural by-products (e.g. sun-
    flower meal), than with a 32% protein manufactured aquaculture
    feed.

–   At the end of the cycle (end of August/September), if stocking densities
    exceed 35,000 shrimp/ha, it is possible to boost growth by switching
    to a protein-enriched feed, if the natural productivity of the pond is not
    sufficient.

–   In more intensive systems (density greater than 4 shrimp/m^2), the daily
    feed ration is calculated by multiplying the estimated shrimp biomass by a
    feeding rate that varies according to the average weight of the shrimp. For
    an average weight of 20g, this rate is around 2%.

4.6. Growth monitoring ⓴

4.6.1. Objective

Estimate the average weight and growth of shrimp in grow-out ponds to identify potential problems, adapt fertilisation and optimise production.

4.6.2. Species or system specificity

M. rosenbergii is a species that lives in turbid waters at the bottom of ponds
and is rarely visible from the shore. It is difficult to assess the growth of shrimp
without catching them. Their growth depends on many factors linked to the
quality of the water and the ecosystem of the ponds; comparing the growth
of the shrimp with reference data enables any anomalies to be detected and
corrected.

4.6.3. Points to watch

Physico-chemical water parameters: temperature, turbidity, dissolved oxygen,
water level.

4.6.4. Indicators of good practice

- Steady, uninterrupted increase in average shrimp weight.
- Growth curve aligned with the reference curve (Figure 4.5).

4.6.5. Instructions for use

4.6.5.1. Weighing frequency

Weighing is carried out as frequently as possible. Weekly is ideal, but difficult to
maintain. Once a fortnight is a good frequency, or at least once a month.

4.6.5.2. Captures

For each rearing pond, use a seine, a net or a bichette (8 or 10 mm mesh) to
catch a sample of at least 20 shrimp. Keep all the prawns, large and small.
The greater the number of shrimp, the more representative the population and
the better the estimate of average weight.

4.6.5.3. Weigh-in

Weighing was carried out using an electronic balance with a precision of 0.1 g. The prawns are drained and placed dry in a tared container. The total weight of the sample is measured, the exact number of prawns is counted and the average weight can be calculated.

It may be a good idea to repeat the operation (capture and weighing) on a new sample and take the average, particularly if the first result obtained seems aberrant.

In addition to the average weight, it is interesting to observe the dispersion of sizes by producing a histogram by size class.

4.6.5.4. Growth monitoring

The data is recorded and compared with the data from the reference growth curve (Figure 4.5).

Plotting all the measurements on a graph allows you to identify an abnormal fall or rise in growth.

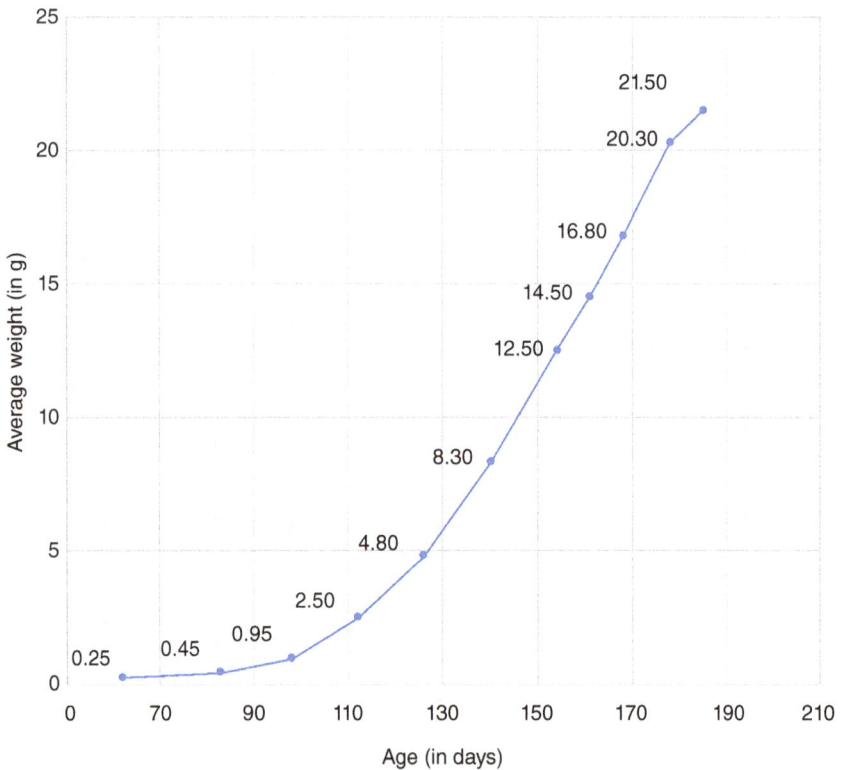

Figure 4.5. Growth curve for reference juveniles. Juveniles aged 60 days (PL60) placed in ponds at the end of May and harvested after 4 months.

4.6.5.5. Interpretation results

The growth curve should be exponential at first and then linear.

If there is a downward trend, the cause must be identified. It may be due to insufficient fertility in the pond and/or stressful physico-chemical parameters in the water: insufficient temperature, lack of oxygen, etc. ⚙. Corrective action should be taken.

4.6.6. Recommendations and special precautions

–   In the case of abundant macrophyte algae, this complicates capture.
–   Take samples from several locations in the basin to ensure that the population is properly represented.
–   From late August or early September, many individuals reach sexual maturity, and the morphotypes of male shrimp can be identified. Estimating the average weight of each morphotype and their proportions in the population provides useful additional information.
–   The survival rate of the shrimp and their precise average weight can only be accurately calculated when the pond is finally emptied.

4.7. Predator prevention ⚙

4.7.1. Objective

Limit production losses caused by predatory animals by implementing preventive measures.

4.7.2. Species or system specificity

M. rosenbergii is a species that prefers turbid water. The turbidity of the water can protect it from numerous predators. Its main defence is speed of escape. However, shrimp remain a prime prey for many predators. Farming losses can be considerable, and predation is a risk to be taken seriously. In Europe, predators include birds (divers and waders), reptiles (snakes, turtles), aquatic insects (dykes, dragonfly larvae) and mammals (mink, otters). It should be noted that shrimps themselves can be predators of their own kind (cannibalism).

4.7.3. Points to watch

–   Connection of basins with natural water.
–   Filling the ponds.
–   Surrounding wildlife habitats (forests, watercourses, etc.).
–   Presence of predatory bird nests in ponds or on the edge of ponds.

4.7.4. Indicators of good practice

–   Good harvest survival rate (over 80%).
–   Anti-bird nets covering all the ponds.
–   Draining carried out before juveniles are introduced.
–   Absence of predators or traces of predators on the farm.
–   Shrimp growth in line with theoretical curve.

4.7.5. Instructions for use

4.7.5.1. Observation of the presence of predators

The presence of predators, particularly birds, can generally be detected by observing the environment. Some predators can arrive in groups over a short period of time and wreak havoc. This is particularly true of the cormorant, a major predator for aquaculture in Europe; it can feed on large prawns and therefore represents a threat. The birds will be easy to see, apart from certain small, very discreet divers such as the great spotted grebe (Figure 4.6).

Figure 4.6. The great spotted grebe, a small, discreet bird that is fond of crustaceans and molluscs and is common in shrimp ponds. It is a protected species.

The presence of predatory insects such as dytids and dragonfly larvae is detected by catching them in nets, for example when sampling shrimps for growth monitoring.

Predatory mammals (mink, otters) are difficult to observe. They can be detected by the tracks they leave behind: droppings, footprints, shrimp corpses.

4.7.5.2. Preventive action

- **Birds:** the most effective preventive measure is the bird protection net. This consists of a net with a mesh of around 15 cm installed above and around the edges of the pond using a system of cables. Simple networks of cords (without netting) stretched over the ponds are less effective but can disturb certain birds such as cormorants. Light and/or sound devices (scarecrows) can be effective, but only for a short period, as the birds get used to them.
- **Insects:** insect larvae pose a risk to young shrimp or juveniles, so the sensitive period is when the shrimp are stocked at the start of the season and in the weeks that follow. Putting large juveniles (0.5 g) into the ponds helps to reduce this predation. It is important to ensure that there is a dry period a few weeks before introducing the juveniles in order to limit the development of insects at the start of the season.
- **Fish:** fish can either come from the surrounding waters or from previous cultures. To prevent the introduction of fish and insect larvae, it is essential to use screens or strainers when filling ponds with water from the natural environment. If the presence of fish is suspected (e.g. trout left over from winter production), a trammel net laid across the pond will allow them to be caught.
- **Mammals:** trapping is generally not an option, as some species such as the otter are protected. Contact the French Office for Biodiversity (OFB) for more information. The most effective, but very costly, form of protection is to enclose all the ponds hermetically with wire mesh, the base of which is buried deep underground.

4.7.6. Recommendations and special precautions

- Some predators are on the IUCN (International Union for Conservation of Nature) red list; killing and capturing them is prohibited. In the event of accidental capture, they must be released off the farm.
- Predators differ depending on the geographical location of the farm. An analysis of the dangers created by predators is important for optimising production.
- Protective bird netting is essential. Many birds are migratory. It is advisable to monitor the ponds on a daily basis to spot any birds trapped in the netting and rescue them.

- The presence of frogs and toads in the ponds generally indicates a low presence of predators.
- Finally, ill-intentioned humans can also pose a threat through theft. The site should preferably be fenced off, and surveillance cameras can be installed. The use of automatically triggered "hunting" cameras will also have the advantage of allowing any nocturnal predators to be seen.

Health Management

5

M. rosenbergii is a species little affected by disease in semi-extensive farming, compared with intensively farmed marine shrimp species. Epidemics caused by viruses caused considerable economic losses in the 1990s in penaeid shrimp farms. The main threats are white spot syndrome *virus* (WSSV) in Asia and Taura disease virus in the Americas (FAO, 1997).

The diseases encountered in *M. rosenbergii* farms are essentially linked to poor management of the rearing environment. The best way to prevent the disease is to maintain a good quality living environment, coupled with effective biosecurity measures.

The Macrobrachium genus is allochthonous in Europe, occurring neither in the wild nor on farms. This makes Europe a disease-free zone for specific Macrobrachium diseases, a particularly attractive and privileged situation that must be preserved through good sanitary practices!

5.1. Farming and biosecurity practices that protect the environment

In aquaculture, farming practices can subject shrimp to stress situations that can affect them. Stress factors include water pollution, handling, oxygen depletion, temperature variations and overcrowding. They can lead directly to death (in the case of anoxia, for example) or encourage the development of opportunistic or infectious pathogens.

The semi-extensive farming method suggested in this guide is characterised by low stocking densities of less than 4 shrimp/m^2 and a feed based mainly on the natural productivity of the ponds during the grow-out phase. These methods are particularly favourable for maintaining good shrimp health.

Low densities minimise stress for shrimp, compared with intensive systems, as well as the risk of pathogens spreading. Natural feeding provides the nutritional

DOI: 10.1079/9781836993162.0005

elements needed to boost the non-specific immunity of crustaceans (they do not have a specific immune system found in vertebrates): immunostimulants such as vitamins and trace elements.

As *M. rosenbergii* is an allochthonous shrimp, the pressure of infection by pathogens in the natural environment is low. Of the pathogens present in Europe, only white spot disease (WSSV), present in other decapods, appears to be a potential threat to *M. rosenbergii*, and the species is not very susceptible to it in the adult state (see Table 12).

The livestock management practices proposed in this book, particularly in Chapter 4, considerably reduce the risk of the appearance or spread of pathogens, and thus constitute direct or indirect preventive measures.

The practice of a seasonal sanitary vacuum, imposed by the geographical location (temperate zone) and carried out every year in winter, is a first preventive measure. The breeding ponds are systematically emptied of shrimp (at harvest time) for a minimum continuous period of 7 months. Even if a few individuals escape the harvest at the end of the summer, they cannot survive temperatures below 13°C and are therefore naturally eliminated from the ponds at the end of October or beginning of November. This rules out the persistence of *M. rosenbergii* or pathogens potentially carried by *M. rosenbergii* in the external environment.

Another major preventive measure is to dry out the ponds before the juveniles are put back in the water in spring. This drying out, which can be combined with liming the soil in the ponds, eliminates aquatic organisms and cleans up the soil in the ponds, considerably reducing the risk of introducing pathogens.

The design of the aquaculture site as a closed facility also appears to be an effective disease prevention measure. This precaution, initially designed to limit the risk of escape of exotic species ④ and ⑧, also limits the risk of contamination *via* natural waters by chemical agents (pesticides, various contaminants, etc.) or biological agents: pathogenic micro-organisms or animal species carrying these pathogens. This is an advantage that will also benefit other species that may be reared alongside shrimp during the winter, particularly trout. If external water is brought in, particularly to compensate for any losses through evaporation, it is important to set up a filtration system (using a grid or strainer) for the water introduced, to avoid the introduction of species that are potential vectors of disease, crayfish in particular and their larvae (avoid bringing in spring water for the latter).

Finally, additional biosafety measures can be taken:

– disinfection of boots at site entrances for workers and visitors;
– workers' working clothes (boots and waders in particular) specific to the workplace;
– introduction of juveniles only from free farms that are regularly screened for the main diseases (white spot disease).

The post-larvae currently available in France come from a batch imported from the United States that has been certified free and actively tested for white spot disease virus (WSSV). Progeny from this healthy batch have been successfully used for production in the Gers since 2018 and have not experienced any pathological events. It is recommended to be extremely vigilant for any new external supply in order to limit the risk of pathogen-carrying shrimp being introduced into France and Europe.

It will be vital to make a collective effort to maintain this favourable health status in order to guarantee the sustainability of this new sector.

5.2. Main causes of mortality and associated measures

The nursery and grow-out phases are distinguished not only by the age of the individuals reared (post-larvae *versus* juveniles and adults), but above all by the rearing environment, above ground in the nursery phase and outdoors in the grow-out phase.

5.2.1 Nursery

In nurseries, post-larvae are concentrated and reared in a closed recirculating system (RAS) with little or no water renewal. The system is dependent on artificial water control techniques: biological filter, aeration and heating.

For example, a power failure can be fatal, causing the sudden death of an entire batch through anoxia linked to a stoppage in aeration. A sudden change in a physico-chemical parameter of the water can also lead to sudden death. Care must be taken with temperature variations when adding large quantities of water or changing the tank of a batch, to avoid thermal shock.

A frequent cause of mortality in nurseries is an increase in nitrite levels due to malfunctioning of the bio-filter, particularly if it has been incorrectly activated when the post-larvae are introduced. The critical threshold is a nitrite level greater than or equal to 3 mg/l. As shown in Figure 5.1, nitrite toxicity can be latent, with mortality only appearing after a period of exposure that can last more than a week.

Figure 5.1. Mortality of post-larvae during the appearance of a nitrite peak in the nursery.

A sudden death is characterised by numerous corpses deposited on the bottom of the tank. If the water is murky, they will not be visible. It is therefore advisable to probe the bottom of nursery tanks daily with a dip net to check for any unusual corpses. A drop in the activity of surviving shrimp is generally observed in the event of significant mortality in a batch; in particular, they feed less or stop feeding, a sign of major discomfort, most often linked to the physico-chemical quality of the water. Appropriate corrective action must then be taken ⑫.

In the event of a high mortality rate in the nursery, it is essential to remove the corpses from the tanks quickly to avoid any deterioration in water quality.

Chronic, progressive mortality can also occur and will be more difficult to detect. Shrimp feed on corpses, leaving little trace of the few losses. However, regular observation of the rearing tanks will reveal the presence of dead shrimp carried by active shrimp.

The causes of progressive mortality can be:

- **Over-density**. As the shrimp grow, it is important to reduce their density, in particular by adding substrate ⑭.
- **Undernourishment**. This can trigger cannibalistic behaviour.
- **Poor water quality**. If water quality parameters remain close to lethal values, progressive mortality may occur ⑫.
- The occurrence of a **disease** (see Tables 12 and 13), of viral origin, for example nodavirus white tail disease, or bacterial.

Illnesses are often linked to bacterial proliferation in water contaminated by organic matter, for example in the event of over-feeding. Radical measures such as cleaning or even emptying and disinfecting nursery tanks may be necessary to put an end to the problem.

5.2.2 In grow-out ponds

In outdoor ponds, mortality is difficult to observe because dead shrimp are generally not visible due to the turbidity of the water. Any dead shrimp can be found by probing the bottom with a dip net.

Mortality can of course be caused by **predation**. In this case, corpses will rarely be seen and it is at the final harvest that a low survival rate will be recorded, indicating regular and invisible losses. Detailed information on this problem can found above ㉑. **Cannibalism** can also be added as a possible cause of mortality and low survival rates; in ponds, it is caused by undernourishment, itself linked to the lack of fertility of the environment.

The most specific cause of mortality in temperate zone breeding is probably the **cold** at the end of the grow-out season. A sudden drop in outside temperature can rapidly lower water temperature, particularly if it is associated with a cold wind spell. Below a water temperature of 16°C, shrimp show a marked drop in activity and the weakest individuals may begin to die, a phenomenon that worsens with a further drop in temperature. Below 13°C, all the shrimp eventually die within a few days. Prevention requires daily consultation of weather forecasts from September onwards, and the ability of the producer to launch an early total harvest if the forecasts are poor.

An episode of sudden mass mortality of shrimp can be due to **anoxia**. This occurs when the environment is asphyxiated by a *bloom of* phytoplankton or a proliferation of cyanobacteria ("blue algae"). An overabundance of phytoplankton or cyanobacteria leads to an overconsumption of oxygen (at night) and an overconsumption of nutrients, which can lead to sudden death as a result of environmental exhaustion. Dead algae increase the quantity of organic matter to be broken down, leading to overconsumption of oxygen and asphyxiation of the environment. The environment becomes potentially lethal for *M. rosenbergii* below a level of 1 mg/l of dissolved oxygen. The shrimp then accumulate on the banks near the surface in search of oxygen and can die off massively, in extreme cases within a few hours. There are a number of warning signs: the Secchi is around 20 cm deep, the water is bright green and green floating masses are forming on the surface (cyanobacteria blooms). The situation becomes dangerous if the water suddenly turns brown, indicating the death of phytoplankton or cyanobacteria. To forestall an accident, you need to measure dissolved oxygen regularly with an oximeter, especially at the end of the night. The best prevention is to activate the aeration of the pool at night (and during the day if necessary).

An **excessive pH** at the end of the day (pH > 9) is also an indicator of an excess of phytoplankton; it can occur on hot, very sunny days. A pH above 9.5 can be lethal.

Good fertilisation control helps prevent phytoplankton *blooms* ⑱ and ⑲.

Another cause of sudden death can be **contamination** by a pollutant: a pesticide, an uncontrolled dump or industrial waste. The design of a closed installation, with no direct contact with natural water, limits this risk of pollution. However, care must be taken when adding water to compensate for losses due to evaporation in mid-summer, and the quality of the water must be checked.

Finally, **disease** can cause mortality in grow-out ponds. However, diseases are rare in adults in semi-extensive environments and are more likely to be linked to poor water quality, leading to the development of opportunistic bacteria or mycosis. Mention should be made of white tail disease, which affects juveniles and can cause significant mortality. However, no disease has yet been detected in pond farming in Europe.

5.2.3 Main diseases of *M. rosenbergii*

The main diseases that can be encountered in semi-extensive livestock farming in temperate zones are presented in Tables 12 and 13, which are not exhaustive.

These diseases are particularly common in South-East Asia, the main breeding area for this species. They are mainly viral or bacterial; there are also fungal or algal diseases.

The contributing factors are common and are characterised by zootechnical imbalances such as over-density or poor water quality.

Table 12. Summary of the most frequent viral diseases of *Macrobrachium rosenbergii*.

Disease	Aetiological agent	Stage reached	Main clinical signs	Presence zone	Remarks
White Tail disease	*Macrobrachium rosenbergii Nodavirus* (MrNV) Associated with *Extra Small Virus* (XSV)	Larva – PL – Juvenile	Body whitening. Mortality possibly reaching 95%.	Potentially worldwide. Observed in the West Indies, China, India, Thailand and Australia.	Specific to *M. rosenbergii*
White spot disease	*White Spot Syndrome Virus* (WSSV)	Adult	None		
		PL - Juvenile	Appearance of white spots on the exoskeleton. Increased mortality and cannibalism.	Potentially worldwide. Observed in the Mediterranean, South-East Asia, Japan, Korea and America.	Regulated disease affecting all decapods.
		Adult	Poor clinical expression.		
Macrobrachium muscle virus	*Macrobrachium muscle virus*	Juvenile	Muscle opacification and progressive necrosis. Mortality up to 50%. Appears within 10 days of the animals being transferred to the grow-out tank.	Observed in Taiwan.	/
Iron Prawn Disease	Suspicion *Flaviviridae*	PL, juvenile, adult	Sexual precocity and stunted growth.	Economic importance in China since 2010, research still ongoing in 2021.	*Iron Prawn Disease*

Table 13. Summary of the most frequent bacterial, fungal and algal diseases of *Macrobrachium rosenbergii*.

Disease	Aetiological agent	Stage reached	Main clinical signs	Presence and/or comment area
Black spot Disease	Vibrio, Pseudomonas, Aeromonas	All Larva - PL - Juvenile - Adult	Melanisation of the carapace, localised or extensive.	Potentially worldwide.
White Post larval Disease	Rickettsia	PL	Bleaching the body of lorries. Mortality up to 95%. Hepatopancreas atrophy.	Potentially worldwide.
Mycoses	Debaryomyces hansenii associated with Metschnikowia bicuspidata	Adult	Yellowish to greyish muscle. High mortality.	Accumulation of organic matter and eutrophication of ponds.
Epibionts	Green algae : Chlorophytes, Oedogonium, Cyanophyte, Lyngbya	Adults, mainly blue claws.	Presence of algae on the surface of the exoskeleton of adults.	Basin not turbid enough, Secchi > 40 cm. Excess of organic matter in suspension.

Harvesting and Marketing

6

6.1. Harvest ㉒

The last phase in the production cycle, shrimp harvesting must be carried out before the cold weather arrives in temperate latitudes. This seasonal harvest is known as "total", which characterises a discontinuous farming method, with the ponds unoccupied by shrimp from early autumn until late spring. However, it may be useful to harvest shrimp earlier to spread out the sales period, which is very short in temperate zones. One or more so-called "selective" harvests can then be carried out a few weeks before the final harvest, if the size of the shrimp permits. Only the largest individuals are harvested, in particular the large blue-clawed males, which will allow the smaller individuals not harvested to catch up with their growth. This makes the most of *M. rosenbergii*'s capacity for 'compensatory growth', with the small dominated animals accelerating their growth when the dominant ones are removed.

6.1.1. Choosing the right time to harvest

In temperate zones, freshwater shrimp are harvested in late summer, when water temperatures drop below 20°C for the daily maximums (pond bottom temperature). The growth of *M. rosenbergii* is greatly slowed down, if not eliminated, below 19 or 20°C. As a result, extending the rearing period will not increase yields, but only provide an opportunity to stagger the harvesting period. As autumn and winter arrive, it will be important to harvest the prawns before they are affected by the cold. It is therefore advisable to carry out the final harvest before the water temperature drops below 17°C.

Below 16°C, the shrimp will be under severe stress and their activity will slow down; they will move slowly and the risk of mortality will increase. Between 13 and 15°C, the weakest shrimp will wither and, below 13°C, the temperature will be lethal for the whole batch. So it's imperative not to wait for

DOI: 10.1079/9781836993162.0006

water temperatures in the ponds to fall below 15°C before harvesting, or risk major production losses.

In the south of France (for example, in the Gers), the latest date for harvesting is generally between 1 and 15 October. However, an exceptionally early cold spell can occur at the end of September, as was the case in 2020, with water temperatures dropping rapidly below 15°C over a few days. Cold, windy weather combined with rain can cause pond temperatures to fall rapidly. It is therefore essential to react quickly and drain the ponds as a matter of urgency to limit losses. This type of decision is made carefully studying the weather forecasts for the moment and for the coming days.

The time of day for harvesting is also important, and its choice will depend on the weather forecast. In early autumn, it is common to have cool mornings (temperatures below 10°C) and warm, sunny afternoons (30°C). When the water level in the ponds is at its lowest, you need to ensure that conditions are favourable for the survival of *M. rosenbergii*. Indeed, if the air temperature is cold (below 10°C) when the pond is at the end of its emptying process, the water temperature can drop very quickly, as there is no longer the thermal protective effect of the water column. The shrimp become lethargic and do not have the strength to move towards the exit of the pond. Many shrimps can then wash up on the bottom of the empty pond and freeze to death.

Conversely, if the end of the emptying process occurs at a warm and sunny time of day, there is a risk of mortality, either through anoxia or for shrimp stranded by direct action of the sun. A shrimp in full sun outside the water will die within a few minutes.

As a general rule, we recommend harvesting during the morning, ensuring that the pond has reached its lowest level (complete emptying) by mid- to late-morning, when air temperatures are suitable. Ideally, harvesting should take place when the outside temperature is around 18-20°C and the weather is overcast.

6.1.2. Harvesting methods

6.1.2.1. Selective fishing

For selective harvesting, the method of choice is fishing with a net: the seine. This is the technique used in tropical areas, such as Guadeloupe, for continuous farming. It can only be applied to ponds with no obstacles, and therefore no substrates.

The seine is a rectangular net bounded by two lines, the selvedges, the surface line being equipped with floats and the bottom line with sinkers (Figure 6.1). Its height corresponds to 1.5 to 2 times the maximum water height of the pond. The recommended height for freshwater shrimp ponds is therefore 2 to 2.5 metres. If possible, the length should cover the entire width of the pond and should therefore be 1.5 to 2 times this width. However, it is possible to use a smaller seine and fish a basin in portions. This is particularly useful for taking

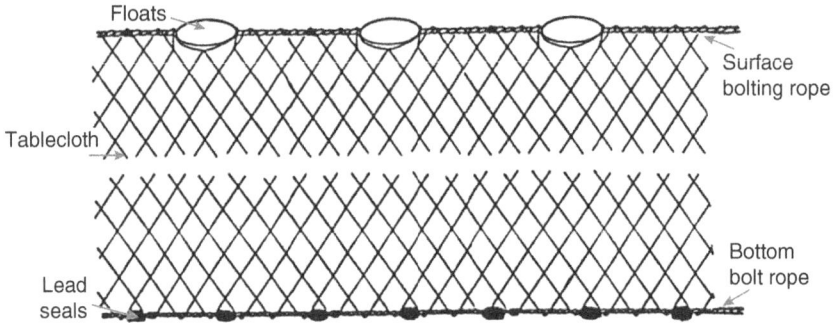

Figure 6.1. General diagram of a seine net.

samples (for weighing, tasting or to start selling small quantities to restaurants, for example).

The mesh size of the seine can be small (10 mm) if you want to catch all the shrimp, including small individuals, or large if you only want to catch large shrimp (20 mm for individuals weighing more than 15 g).

The method is similar to pond fishing techniques. Several people are needed, depending on the size of the seine. Two people (or more) pull each end of the seine and one or more watch over the net to ensure that it is moving properly. In particular, you need to use your foot to check that the bolt rope at the bottom of the net is in the right place and is not lifting. As prawns are benthic animals, they easily escape under the net. You need to choose a suitable spot on the bank for hauling up the net: gently sloping and clear.

This technique, especially if used on parts of a pond, only allows partial harvesting of the shrimp. Even after several passes, the number of shrimp remaining in the pond can be high. Final harvesting by total emptying is therefore essential at the end of the season.

6.1.2.2. Final fishing by emptying

This major operation is a delicate one. It is the last stage in the rearing process and must be carried out carefully and methodically to avoid losses. Avoid crushing the shrimp with your feet or in the landing nets. The stress generated by sudden changes in temperature or careless handling will reduce the shrimp's chances of survival. If shrimp are stored in trays after harvesting, this can also affect the quality of the shrimp (risk of accelerated spoilage after killing). Temperature, as well as dissolved oxygen levels, need to be monitored and controlled during emptying operations.

Grow-out ponds must be designed to allow for this total emptying ⑦. Ideally, they should be equipped with an external fishery, made of concrete or any other system (net fishery possible). For ponds not equipped with an external fishery, harvesting will take place in a capture pit, an area dug out at the level of the pond emptying system (Figure 6.2).

In both cases, the steps involved in a total drain are as follows:

- Anticipate the emptying date, as described above, depending on the weather forecast.
- Gradually start emptying one or two days before the set harvest date, depending on your knowledge of how quickly the pond will empty and the weather. Boards should be removed if the emptying system is a monk, or the stanchion lowered if it is a tilting stanchion system. Some ponds not equipped with a draining system can also be drained using a pump.
- During the night before harvesting, lower the water level in the pond to 30 or 40 cm (at the low point outside the capture pit) and install an aeration system if there is any doubt about the oxygen level.
- Fully open the drainage system at the end of the night or very early in the morning.
- Harvesting during the morning. A team of several people is needed. First, collect any shrimp that may have washed up on the bottom of the pond. Then collect the shrimp using a dip net in the catching pit (Figure 6.2) or in the fishery if there is one (Figure 6.3).

It is advisable not to walk in the water during emptying to avoid leaving footprints at the bottom of the pond; shrimp can take refuge in these holes, which complicates harvesting. For systems with an outside fishery, shrimp tend to accumulate in the emptying area and leave the pond at the same time when

Figure 6.2. Emptying a pond with a capture pit.

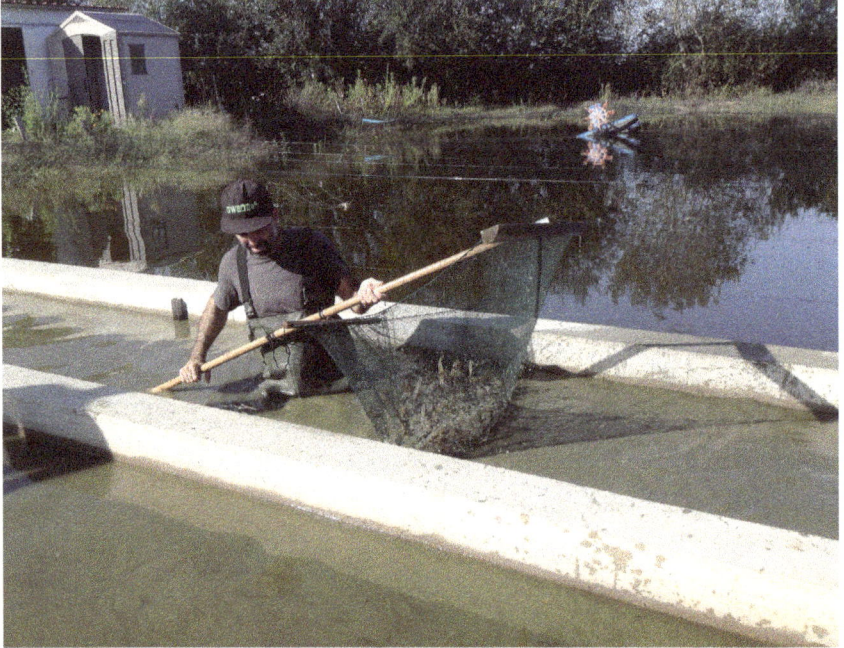

Figure 6.3. Net harvesting in an outdoor fishery.

the water level in the pond is at its lowest. It is advisable to wait and let the prawns emerge naturally without intervening, to avoid panic and exhaustion.

The shrimp are collected in baskets (of the market gardener's harvesting basket type, Figure 6.4); these baskets can remain partially immersed during harvesting. If the shrimp are very dirty, for example if they have been collected from a muddy capture pit, they can be rinsed in a first tank of clean water by immersing them for a few minutes. They are then transferred to well-ventilated tanks to drain. This phase allows the shrimp to swim and get rid of impurities inserted in the gills or on the rest of the body. They remain there for a few minutes to a few hours before being slaughtered.

Weighing will be carried out before or after this clean water bath, or even after slaughter. The calculation of the total number of shrimp harvested, necessary for monitoring the performance of the farm, will be based on the total harvest weight, divided by the average individual weight of the shrimp. The latter will be determined by weighing several samples taken at different times during the harvest.

6.1.2.3. Post-harvest live shrimp storage 23

Live shrimp may need to be stored overnight or for several days. This type of storage allows the shrimp to be sold live, but above all makes it easier to sell a harvest over several days and to carry out slaughter operations in several stages.

Figure 6.4. Shrimp in the harvesting basket.

Stored shrimp must have been handled as gently as possible during harvesting if a good survival rate is to be achieved during storage. *M. rosenbergii* does not cope well with high densities and there is a high risk of cannibalism or mortality caused by fighting. A good survival rate is obtained for a stocking of 5 kg per m³ on the condition that the shrimp are provided with abundant support (substrate). If the nursery is built close to the grow-out ponds, the nursery tanks and their substrate can be advantageously used for this post-harvest storage.

The storage temperature must be relatively low to limit shrimp activity and therefore the risk of injury and mortality due to fighting. A temperature of 18 to 22°C is ideal. You will need to aerate the tanks thoroughly and change the water if they are stored for more than a few days. Shrimp should not be kept for more than a week.

6.1.2.4. Expected return

Production data from temperate zones in the United States show a yield per hectare of 800 to 1,000 kg in a semi-extensive, low-input system with a stocking rate of 3.5 shrimp/m² ❸ . The average weight of individual shrimp is over 30g. This yield is the maximum level that it seems possible to achieve in southern Europe with the semi-extensive model as described in this best practice guide. Achieving such a yield requires high summer temperatures, good control of fertilisation and a high survival rate of around.

Several years of production in the Gers have yielded 650 kg/ha at best, with an average weight of 25 grams. However, some basins are less efficient due to high predatory pressure or poorly controlled fertilisation.

For example, it is currently reasonable to expect a yield of between 500 and 600 kg per hectare in south-western France.

6.2. Slaughter and cold storage ㉔

6.2.1. Alteration and shelf life

Flavour and texture are the two main criteria for the organoleptic quality of fish products and shrimps in particular. Alteration of texture is a major determinant of the quality of *M. rosenbergii* prawns. It results in a softening of the flesh that is detrimental to its quality: the flesh loses its firmness and crunchy texture, which are so appreciable when it is fresh.

Crustaceans spoil more quickly than fish, due to their high content of small metabolites such as free amino acids. The latter also contribute to their subtle flavour.

Spoilage can be caused by bacteria, naturally present in the digestive system and on the surface of the shell; they penetrate the cells and multiply following the death of the shrimp. Spoilage can also be caused by the shrimp's digestive enzymes, particularly those from the hepatopancreas, which act on the shrimp flesh when the animal dies. Shrimp spoilage starts at the base of the tail (or abdomen anatomically), in contact with the head, and then spreads to the rest of the tail.

The enzymatic blackening of the carapace, also known as "melanisation", can also be observed in *M. rosenbergii*, but is less rapid than in shrimps. It only appears after several days in a cool place, which avoids the need for treatment with bisulphite, an allergenic substance.

The care taken during harvesting, the slaughtering technique and the speed of the cold treatment will have a major influence on the spoilage of the prawns.

When fishing, care must be taken not to let the prawns asphyxiate out of the water, to avoid contamination by mud and crushing, and to rinse them thoroughly.

Following slaughter, rapid cold treatment is the only effective way to combat spoilage, which begins immediately after the animal's death. Spoilage is accelerated by temperature; it occurs within a few hours at room temperature.

6.2.2. Felling technique

At the time of harvesting, the live shrimp are at room temperature, the same temperature as the rearing water. Ifremer (1991) considers the slaughter method using an ice water bath to be the most effective. This is the method used by the Parc Aquacole (Guadeloupe) and Gascogne Aquaculture (Gers).

This method achieves two objectives in a single operation:

– the rapid killing of shrimp (slaughter), limiting animal suffering as much as possible;
– rapid cooling of killed shrimp to reach a core temperature of 2°C or less in just a few minutes, thereby limiting spoilage.

6.2.2.1. Ice bath procedure

One part clean water and one and a half parts ice are mixed in a tub. The prawns are immersed in the water for 30 minutes, the time needed for them to cool to the core. During this time, make sure there is always unmelted ice in the tank, to ensure a temperature close to 0°C. For example, for 10 kg of prawns, you would use 10 litres of water and 15 kg of ice.

A production diagram illustrating this method is shown in Figure 6.5.

Slaughter and subsequent handling (sorting, packaging, cold storage) are carried out in a workshop operating in accordance with current health standards (European Hygiene Package regulations). The premises, equipment and materials must be designed to avoid any risk of food contamination. Walls, floors and the cold room are made of suitable, washable and smooth materials. Work tables are made of stainless steel. Good hygiene practices must be applied, in particular a cleaning and disinfection plan for the premises and equipment.

Packaging: polystyrene insulated containers

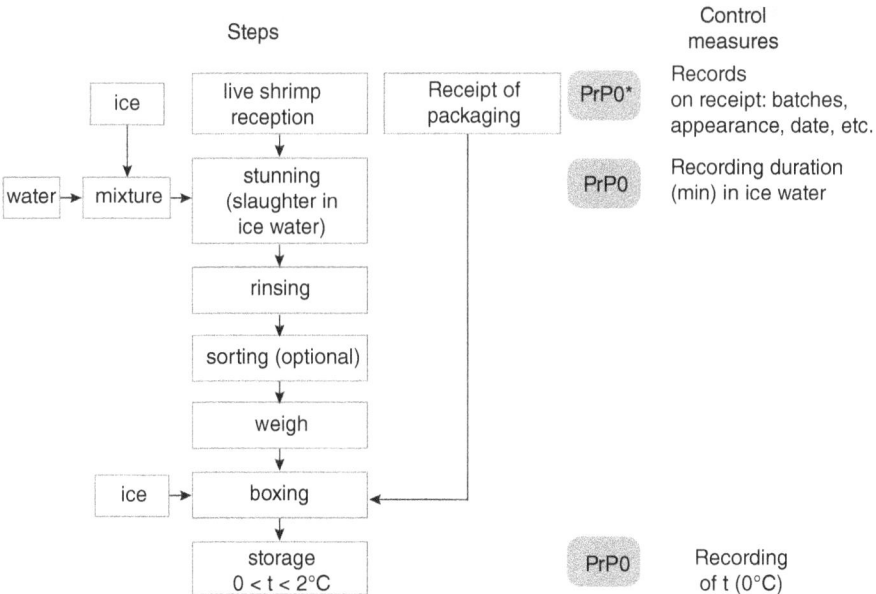

Figure 6.5. Diagram of the shrimp slaughtering method in an ice bath. * PrPO: Pre-requisite Operational Programmes.

In the case of direct delivery or local sales to intermediaries who themselves carry out direct delivery, it is possible to carry out these operations in a workshop, in derogation of the health approval. In this case, the regulatory requirements are less stringent than for EC-approved workshops. You should contact the department in your county responsible for food hygiene (DDPP) to find out about the authorisation procedures.

6.2.3. Packaging and cold storage

Following slaughter, manual sorting can be carried out, followed by weighing (Figure 6.6). This sorting is useful for separating small shrimp (in particular light-clawed males). Several size categories can be marketed depending on the harvest and customer demand For catering, it is preferable to supply batches of uniform size.

Fresh prawns are packaged in bulk in perforated food containers. The prawns are placed in thin layers and covered with ice; it is advisable to avoid direct contact between the ice and the prawns by using cling film. The trays filled with shrimp on ice are placed directly into cold storage or can be inserted into reusable isothermal boxes for shipping. These trays and reusable boxes must be carefully disinfected after use.

Figure 6.6. Sorting and weighing freshwater shrimp in a dedicated workshop.

The use of disposable polystyrene isothermal boxes, with openings in the bottom to allow water from melting ice to drain away, is widespread in fish shops, but is not recommended because of its environmental impact (polystyrene is a pollutant). If the choice is to use single-use packaging, it is possible to replace polystyrene with waterproof cardboard boxes. Specialist packaging companies are now offering high-performance models. These more environmentally-friendly alternatives will certainly be the norm in the near future.

If harvesting, slaughtering and cold processing are carried out with care, shrimp will have optimum shelf life. European regulations stipulate a storage temperature of between 0 and 2°C for fishery and aquaculture products.

Shrimp kept on ice at 2°C can be kept for 3 to 4 days without deterioration, and up to 8 days according to some authors (New *et al.*, 2010).

Prawns frozen at will retain their flavour for several months. To be authorised to market frozen products, you need to use a rapid freezing cell, which requires additional investment and technical expertise. These techniques are not detailed in this best practice guide, which has chosen to promote the sale of ultra-fresh prawns, a product that stands out for its quality. Frozen shrimps, which are very present on the imported shrimp market, do not correspond to the logic of short local circuits promoted by this manual.

However, consumers are free to freeze fresh freshwater prawns at home in order to defer their consumption. Producers can advise them on how to optimise this operation. Freezing should be carried out very quickly, as soon as you get home, and only with ultra-fresh shrimp (sold on ice less than 48 hours after slaughter). The prawns should be spread out on a grid or other support in the freezer, to prevent them sticking together and forming a block. Shrimp harvested and slaughtered in early autumn will retain all their quality and flavour if you follow these instructions, so you can enjoy them at the end of the year!

6.3. Marketing

Freshwater prawns are a new consumer product in mainland France and Europe. Sold live or ultra-fresh, freshwater prawns stand out from the marine prawns known to the general public, which are imported frozen or cooked. However, the proposed production model requires seasonal sales over a short period. It will be important for producers to study their market carefully before launching their project. Short and local circuits are niches for which there is a real lucrative market.

6.3.1. A little-known product for European consumers

Apart from certain communities (West Indian, Chinese, etc.) that are culturally fond of freshwater prawns, or travellers who have come across this species abroad, *M. rosenbergii* is unknown to most Europeans. This is quite normal, as the genus Macrobrachium does not occur naturally in Europe and freshwater prawns are not part of traditional diets. On the other hand, Europeans are familiar with marine shrimp, and there is also a tradition of eating crayfish. The consumption of imported marine shrimp has exploded in recent decades since the industrialisation of production in China, South-East Asia and Central and South America. Every year, France imports almost 120,000 tonnes of shrimp for its domestic market, which is almost 3 times more than it consumed before the 1990s. The product has become commonplace and local outlets are impressive.

However, production in mainland France is very limited. There are two "French origin" species on the French market: shrimp bouquet (*Palaemon serratus*) from fishing and shrimp impériale (*Penaeus japonicus*) from farming, with annual production of 350 and 50 tonnes respectively. This represents just 0.3 to 0.4% of the shrimp consumed by the French (source: FranceAgriMer, 2017)!

Shrimps imported into France are mainly farmed shrimps from the peneidae family. Whether cooked or frozen, they are always treated with bisulphites to prevent melanisation. Their production, whether through intensive farming or fishing, is widely criticised by numerous NGOs (e.g. the WWF) and by a growing number of consumers who denounce the environmental damage caused. The destruction of mangroves associated with the construction of breeding ponds has a major negative impact on natural environments, as mangroves are home to ecosystems that are essential to biodiversity. The product is also criticised for its chemical contamination (antibiotics, bisulphites, pesticides, etc.), its lack of freshness and its carbon footprint. To these we can add the over-exploitation of natural aquatic environments through fishing or the use of forage fish in its feed, and the growing problem of animal welfare.

These criticisms are all arguments in favour of eco-responsible shrimp production. Organic or red label semi-intensively farmed penaeid prawns are being developed, particularly in Madagascar. In France, semi-extensive farming of imperial shrimp (*Penaeus japonicus*), established since the 1990s in a few farms on the Atlantic coast, is particularly virtuous (Blachier, CREAA, 1998).

The semi-extensive model of local freshwater shrimp production proposed in this guide is in itself a major selling point and corresponds to current consumer expectations.

The arguments in favour of the proposed model can be listed as follows:

– organoleptic quality of the product: ultra-fresh and natural, with no preservatives or treatments, and a very subtle flavour;
– local production for European consumers;
– limited use of natural resources, in particular feeding on by-products of local agriculture, without adding meal or oil from wild fish;
– absence of discharges into the environment and pollution;
– preservation or creation of bodies of water favourable to biodiversity (aquatic invertebrates, vegetation, etc.), the semi-extensive production system being based on the natural ecosystem;

– transparency of farming techniques, with consumers able to visit the production site;
– animal welfare: low stocking densities, natural habitat and feed, humane killing.

6.3.2. Local short circuits for economic viability

A short distribution channel is one in which there is no more than one intermediary between the producer and the consumer. If there is no intermediary, it is referred to as "direct sale" by the producer. For a local channel, it is the geographical distance between producer and consumer that is taken into account, with marketing taking place within a restricted radius of the production site.

6.3.2.1. Advantages of selling through local short distribution channels

There are many advantages to selling through short, local channels, provided of course that there is a local outlet, which is the case for shrimp in France and Europe.

Firstly, these marketing methods enable us to market an ultra-fresh, even live product, because the time between harvest and sale is reduced to a minimum: there are no or very few intermediaries, transport and distribution logistics.

Consumers are very sensitive to the freshness of crustaceans, which have a short shelf life and whose organoleptic quality is particularly correlated with freshness. Freshness is the key criterion that distinguishes locally produced shrimps from imported shrimps, which are never sold fresh (defrosted at best).

The other advantage of the local short circuit for the producer is a high selling price, justified by the quality of the product. The profits from production go mainly to the producer and are not shared with multiple intermediaries. Transport and storage costs are also kept to a minimum.

Lastly, these marketing methods meet the growing expectations of consumers who want to be reassured about the origin and ethics of the product, and who are aware of the importance of the social ties they generate.

For these reasons, the preferred marketing practices for freshwater shrimp produced in a semi-extensive system are direct sales by the producer and/or sales to local intermediaries who themselves sell directly. These intermediaries may be restaurants, caterers or fishmongers.

These marketing channels enable shrimp farmers to sell their small-scale, non-industrial production while earning a decent income. They contribute to the economic viability of the proposed system, without the need for production subsidies, in particular CAP aid.

6.3.3. Long-distance sales

So-called "long" marketing channels, involving multiple intermediaries between the producer and the consumer, are in practice reserved for high-volume production, and are therefore not suitable for freshwater shrimp produced in Europe. These long distribution channels exist for imported shrimp, but suffer from one major limitation: the impossibility of marketing fresh shrimp. The profitability for the producer of selling *M. rosenbergii* through long distribution channels would be uncertain, particularly if the product had to compete with imported products at very low prices. Long distribution channels imply a low selling price for the producer (compared with short distribution channels) and therefore the need to produce in large quantities to make the farm profitable.

However, there could be a role for an intermediate marketing method based on proximity with *M. rosenbergii*. This would involve one or more operators distributing to professional retailers, restaurants and fishmongers, within a radius close to the production area. If limited distances and logistics enable a shorter post-harvest sales period, and therefore the marketing of ultra-fresh products considered to be 'local' (compared with imported products), then 'long' marketing methods could be tested. Their success will depend on the presence, close to the production areas, of large urban centres with an abundant market, and on targeting customers with high purchasing power.

6.3.4. Product presentation for sale

6.3.4.1. Live shrimp

Some customers prefer to buy the prawns live, the ultimate guarantee of freshness, and do the killing themselves, usually by soaking them in boiling water.

In this way, live shrimp can be marketed, particularly when the sale takes place on the production site, as "farm gate" sales. The shrimp harvested must then be kept alive in above-ground storage tanks [23]. At the time of sale, they can remain out of the water for a few moments (one to two hours) if they are kept in the shade. However, growing public awareness of animal welfare is raising questions about the future of these practices. Many consumers prefer to buy shrimps that have already been killed and cooled, but ultra-fresh of course!

Some professionals, particularly in Asian restaurants, may ask for live shrimp. Live shrimp can also be delivered; this requires transport in water and the use of oxygenated transport tanks, similar to those used for transporting live fish. In this case, the water must be at a temperature of around 20 to 22°C and have an excellent oxygen level (> 5 mg/l).

6.3.4.2. Ultra-fresh prawns

The term "ultra-fresh" refers to fresh food products that have a short use-by date (a few days) and need to be stored cold (between 0 and 2°C for fish products). The distribution circuit is short and fast, taking less than a day.

Ultra-fresh prawns are the form of sale best suited to the production model presented in this best practice guide. It allows efficient and practical management of the product, which is kept at 0-2°C in cold storage and retains all its qualities for a short period of time, from 3 to 4 days. It is advisable to sell them within 24 hours (maximum 48 hours) of slaughter and placing in ice, in order to offer a top-quality, "ultra-fresh" product.

After slaughter, the prawns are kept in cold storage in bulk crates, covered with a layer of ice ㉔. For retail sale, the prawns are weighed using a calibrated scale and delivered in disposable bags or reusable containers brought in by customers.

Products can be sold on the farm, at the local market, at fairs and shows, on the road, in pre-organised sales (baskets), and also at collective sales outlets. The selling price must be clearly displayed, as well as the commercial name of the product, "tropical freshwater shrimp". A price of € 35/kg inc. VAT has been successfully applied in the Gers in 2021 by Gascogne Aquaculture for sales at the production site.

It is essential to respect the cold chain if they are to be transported. They are placed in insulated boxes under a layer of ice. Ideally, they should be transported in a refrigerated vehicle. However, a tolerance is accepted in a non-refrigerated vehicle over short distances, if the products are kept in iso-thermal boxes at the regulation temperature (between 0 and 2°C).

In the same way, prawns can be delivered ultra-fresh to professionals (fishmongers, restaurants). The selling price will then be adapted to the professionals.

6.3.4.3. Cooked prawns

Prawns can be showcased by cooking them for tastings or small-scale catering. Regulations allow fish farmers to use their own products for catering, provided of course that they comply with minimum hygiene standards. Tastings are a great way of attracting the public, and are particularly recommended for new farms. Consumers are unfamiliar with this product, which is not part of their daily lives, so it's a good idea to give them a taste before they buy!

The simplest recipe is probably a la plancha (Figure 6.7): heat a plancha with olive oil, place the whole prawns on each side for a few minutes (they will turn pink), but don't overcook! Add coarse salt at the end of cooking and serve. Simple and effective, plates with 200 or 300 g of prawns can be sold, accompanied by a glass of dry white wine, preferably local, or a craft beer. It is preferable not to add any spices so that consumers can appreciate the taste of the natural product. There are many recipes available, including marinades, flambées, etc., and chefs from local restaurants will be happy to give you a try.

6.3.5. Marketing strategies

As we have seen, the preferred marketing channel is direct sales or sales to re-tail intermediaries (restaurants, fishmongers). As the product is new, it needs to be promoted and actively communicated.

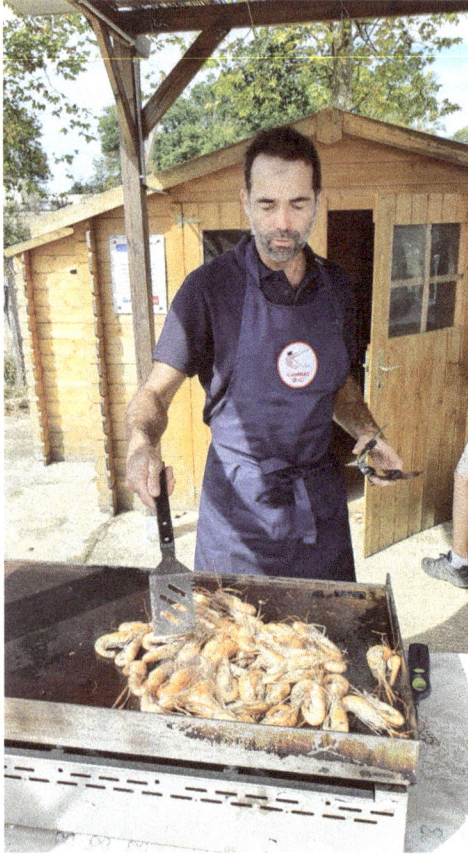

Figure 6.7. Preparation of shrimp a la plancha for tasting on the farm.

6.3.5.1. Communication

Sales must be supported by effective communication. It is all the more delicate because it is seasonal and takes place over a very short period. So it's important not to make any mistakes and to think ahead. It's out of the question to start communicating at harvest time - you'll be too busy!

Communication can be based on the following tools:

6.3.5.1.1. ROADSIDE SIGNAGE   If the farm has the advantage of being located next to a road with heavy traffic.

6.3.5.1.2. POSTERS AND FLYERS   They will be placed in local shops, partner restaurants, tourist offices, local producers, etc. In addition to the cost of production, these tools need to be printed in advance, which can be restrictive if you want to advertise an event (harvest and sale) while retaining flexibility on the dates (depending on the weather).

6.3.5.1.3. SENDING MESSAGES BY E-MAIL OR SMS   This technique requires a mailing list that builds up over the years. It is very effective for announcing sales a few days in advance.

6.3.5.1.4. SOCIAL NETWORKS   Today, this is an essential tool; there are many different apps and their use varies according to age group. The most effective are Facebook and Instagram. A Facebook page can be set up and the producer will need to keep it alive by posting regular information and informing people of sales dates.

6.3.5.1.5. WEBSITE   Consumers will not spontaneously go to a website unless they are specifically looking for information. Such a site must therefore exist. If the number of producers allows it, it would be ideal to share a common site for several producers.

6.3.5.1.6. PRINT OR ONLINE PRESS   The local press will be targeted; most newspapers have paid advertising boxes in which sales or events can be announced. A journalist may also be interested in publishing an article about your production initiative, in which case you should highlight the sales opportunities.

6.3.5.1.7. PARTICIPATION IN TRADE FAIRS AND EXHIBITIONS   This is a useful way of making yourself known outside the sales period and building up a list of customers. However, the frustration will be that you won't be able to show off the product, because it's only available for a short time and you won't be free at that time to take part in trade fairs!

6.3.5.2. Events

The main sales strategy is to sell directly from the production site at harvest time. To attract the public, it is effective to **create events.** Events are easy to publicise on the internet (social networks), in the press, *via* tourist offices... and the public are looking for weekend activities, especially when it comes to gastronomy! The sale of local prawns is an event in itself, if it's infrequent, but it's a good idea to add activities to it.

The aim is to attract customers and make the buying experience a pleasant one:

6.3.5.2.1. OPEN DAYS OR WEEKENDS WITH GUIDED TOURS   Consumers particularly appreciate this approach of transparency and education; it reinforces their confidence in your product and their desire to return.

6.3.5.2.2. CULINARY DEMONSTRATIONS BY A CHEF   This is a great way to showcase a restaurant owner and their establishment, as well as highlighting the product.

6.3.5.2.3. SMALL-SCALE CATERING   Provide tables and chairs for entertaining. Plates of prawns a la plancha are ideal. It's a good idea to invite a mobile restaurant (pizzas or similar) to complete the meal.

6.3.5.2.4. SMALL MARKET OF LOCAL PRODUCERS Inviting other producers (winemakers, cheese producers, market gardeners, bakers, etc.) to sell their products and advertising it allows customers to top up their purchases; producers can activate their own customer networks and bring in the public!

6.3.5.2.5. VARIOUS ACTIVITIES FOR CHILDREN Angling, shrimp workshop, games...

6.3.5.2.6. EVENTS OUTSIDE THE PRODUCTION SITE Gambas evenings at a partner restaurant, with the producer present.

6.3.5.3. Use of official quality signs and/or a collective brand

In France and Europe, official logos are used to identify products that have been awarded an official quality and origin mark. These signs are an official guarantee of quality for consumers. They can guarantee origin (AOC, PDO or PGI), superior quality (Label Rouge), or respect for the environment (Organic Farming: AB). Certain signs could be a convincing marketing argument for freshwater shrimp reared in a semi-extensive system, to attract customers and build loyalty. For the moment, however, there is no question of considering collective membership of a quality mark without the existence of a collective of producers. The AB (or "organic") label, on the other hand, can be applied for individually, provided that the AB specifications for shrimp production are met. The genus Macrobrachium is included, and the semi-extensive farming methods presented in this guide to good practice correspond to the technical specifications of the AB specification. These require in particular:

- basins with natural clay bottoms;
- a density of less than 22 shrimp/m² during seeding, the proposed model being much lower, with 3 or 4 shrimp/m²;
- a diet low in fishmeal;
- a limited number of veterinary treatments.

AB certification can be a real asset for marketing in urban and peri-urban areas, reaching consumers who are disconnected from production. In rural areas, experience in the Gers shows that most consumers are indifferent to this guarantee of quality as long as they buy from the producer and can verify for themselves the eco-responsible nature of shrimp production.

The creation of a collective brand specific to *M. rosenbergii* and guaranteeing a local production ethic would ultimately be an effective way of familiarising consumers with this new product. In particular, it would enable them to identify this ultra-fresh, local freshwater prawn and distinguish it from other prawns on the market. The "Gambas d'ici" brand (Figure 6.8), used by Gascogne Aquaculture since its start-up, could be adopted by other producers who share its interest. A set of virtuous, consensual specifications would then have to be finalised.

6.4. Gross margin calculation

Based on assumptions of yield per hectare, it is possible to calculate the gross margin that could be generated by freshwater shrimp production in Europe's temperate zone. The calculation proposed in Table 14 is based on production and sales data from the Gers (France), under a semi-extensive system fed on sunflower meal.

The gross margin calculation only takes into account the variable costs directly linked to production, with the exception of labour, which is not taken into account, and does not consider the depreciation of the investment.

Figure 6.8. Gambas d'ici brand logo.

Table 14. Calculation of gross margin per hectare of *M. rosenbergii* production in a semi-extensive system (35,000 PLs per ha).

Shrimp sales over 6 months of activity	Variable costs (excluding labour) per ha	Gross margin per ha
– Yield: 500 to 650 kg/ha	– Post-larvae (at €100 per 1,000): €4,000/ha,	Gross margin = sales - variable costs
– Selling price: €25/kg (professional sales) to €40/kg (direct sales)	– Electricity (nursery, aerators, pumps, cold room): €1,500/ha/year,	Low hypothesis: €5,500/ha
– Turnover:	– Feed/fertiliser/lime: €1,000/ha,	High hypothesis: €19,000/ha
– Low hypothesis: €12,500/ha	– Miscellaneous: maintenance and supplies: €	**Estimated average gross margin: €12,250/ha/year**
– High hypothesis: €26,000/ha	**Total: €7,000/year**	
Estimated average turnover: €19,250/ha/year	(Note: PLs account for more than 50% of expenses)	
N.B.: possible yield of 1,000 kg/ha in the United States (Kentucky)		

The main cost is the purchase of postlarvae. For a stocking of 30,000 to 35,000 juveniles/ha and taking into account losses during the nursery phase (around 20%), the requirement amounts to 40,000 PLs/ha.

The calculation assumes that the entire production is sold fresh and through short distribution channels, either direct to the consumer or to local intermediaries (fishmongers, restaurants), at a minimum price of €25/kg excluding VAT. Combining these two sales channels with a yield of between 500 and 650 kg/ha gives an average gross margin of €12,250/ha.

The economic viability of a farm producing only freshwater shrimp on a summer cycle has not yet been demonstrated in Europe. The high cost of building and equipping new ponds can be an obstacle to economic viability because of the high depreciation charges generated. However, the economic potential can be improved if shrimp production is combined with other production in winter, in order to make the ponds, which are available for around 7 months, profitable. One or more cold-water species can occupy the ponds in winter. Trout grow-out is a good example; the margin generated can equal that of shrimp, and the product will benefit from the same short sales channels. Another possibility is to use these ponds for stocking pond fish (carp, for example), as part of a strategy to optimise fish production, as is the case with the nursery ponds in the Dombes region. The presence of another crop in winter will also have the added benefit of fertilising the ponds.

In this way, shrimp production can be seen as complementary to other uses for the ponds (storage of fry, water reservoir for irrigation, fishing, tourism, etc.) - any combination can be imagined!

Conclusion

At a time of ecological transition, when economic players and consumers are expected to rethink production and consumption models, freshwater shrimp farming is a concrete example of food production that meets these challenges. Producing *M. rosenbergii* in a way that is both virtuous and local is part of the drive for European food sovereignty. Freshwater prawns have different characteristics to the imported products (marine prawns) to which consumers have become accustomed over recent decades. *M. rosenbergii* is different because it is farmed in freshwater, and a little restrictive because it is seasonal and unavailable all year round, but it has undeniable advantages: ultra-fresh, tasty, and above all in harmony with our environment and natural resources. Producing it in Europe would also help to restore the social dimension to farming.

The book is designed to keep pace with the environmental challenges of our time. The latest IPCC report (2021) predicts that by 2050 temperatures in Europe will have risen by between 1.5 and 2°C compared with the pre-industrial era. Two approaches are commonly put forward to combat climate change: mitigation measures, aimed at stabilising the concentration of greenhouse gases in the atmosphere, and adaptation measures, limiting the negative impacts or maximising the positive effects of global warming. The proposed model responds to both approaches. It seeks to reduce the use of fossil fuels through extensive local shrimp production. It helps to maintain water surfaces capable of sequestering carbon (phytoplankton). Based on the aquatic ecosystem, it contributes to the conservation and sustainable use of biological diversity. Finally, the cultivation of *M. rosenbergii*, a species resistant to high temperatures, will benefit from the foreseeable rise in temperatures through an extended production period and improved yields.

For these reasons, the *M. rosenbergii* species and the semi-extensive pond production method developed in this guide are in line with the recent strategic guidelines proposed by the European Commission (2021) for more sustainable and competitive aquaculture in the European Union.

It is hoped that this guide to good practice will help new producers of *M. rosenbergii* to get started in Europe; it will help them to design their project and take their first steps in pond management. However, this first version is incomplete, as it is based on limited production experience in Europe. It is to be hoped that other pioneering shrimp farmers will embark on this adventure in the near future, thereby enriching and completing our knowledge of freshwater shrimp farming in Europe.

Bibliography

Blachier P., 1998. L'élevage de la"crevette impériale", version n° 2. Guides techniques du CREAA. 28 p.

Cormoreche J.-C., Bernard S., Chuzeville N., 2014. *Guide des bonnes pratiques en production piscicole en Dombes*. Published by Espace Copie, Bourg-en-Bresse.

D'Abramo L. R., Ohs C. L., Fondren M. W., Steeby J. A., Posadas B. C., 2003. Culture of Freshwater Prawns in Temperate Climates: Management Practices and Economics. Bulletin 1138, Mississippi State University, 23 p.

European Commission, 2021. Strategic guidelines for more sustainable and competitive aquaculture in the European Union for the period 2021-2030. Communication from the Commission to the European Parliament, the Council, the European Economic and Social Committee and the Committee of the Regions. COM(2021) 236 final, 20 p.

FAO, 1997. Review of the state of world aquaculture. FAO Fisheries Circular. N° 886, Rev.1. Rome, FAO. 163 p.

FranceAgriMer, 2017. The shrimp market in France. GEM & 2F Conseil for FranceAgriMer, 96 p.

Griessinger J.-M., Lacroix D., Gondouin P., 1991. *L'élevage de la crevette tropicale d'eau douce*. Ifremer, 372 p.

Guézou M., 2021. *Macrobrachium rosenbergii*: presentation of the species and comparison of growth in recirculated and biofloc systems at different densities. Thesis for the state diploma of veterinary doctor. Oniris - École Nationale Vétérinaire de Nantes. 109 p.

Intergovernmental Panel on Climate Change (IPCC), 2021. Climate Change 2021: The Physical Science Basis, Summary for Policymakers. Working Group I contribution to the Sixth Assessment Report. Printed October 2021 by the IPCC, Switzerland, 31 p.

Lautraite A., Borde G., 2004. Guide de bonnes pratiques sanitaires en élevages piscicoles. Published by the Fédération française d'aquaculture.

Laval G., 2017. Experimental rearing of tropical freshwater shrimp (*Macrobrachium rosenbergii*). Experimental report year 2017. In collaboration with joint research unit INRAE/Oniris, Nantes. 19 p.

New M. B., 2002. *Farming freshwater prawns. A manual for the culture of the giant river prawn (Macrobrachium rosenbergii)*. FAO Fisheries Technical Paper 428. 212 p.

New M. B., Valenti W. C., Tidwell J. H., D'Abramo L. R., Kutty M. N., 2010. Freshwater Prawns: Biology and Farming. John Wiley & Sons Ltd, 551 p.

Quero M., 2021. Development of a good practice guide for rearing a tropical freshwater shrimp, *Macrobrachium rosenbergii*, in temperate zones. Final dissertation. Agrocampus Ouest, Rennes. 59 p.

Schlumberger O., Girard P., 2013. *Mémento de pisciculture en étang*. 5ᵉ edition. Éditions Quae, 222 p.

Tidwell J. H., Coyle S., Durborow R., M., Dasgupta S., Wurts W. A., Wynne F., Bright L. A., VanArnum A., 2002. KSU prawn production manual. Kentucky State University Aquaculture Program, 45 p.

Tidwell J. H., D'Abramo L. R., Coyle S. D., Yasharian D., 2005. Overview of recent research and development in temperate culture of the freshwater prawn (*Macrobrachium rosenbergii* De Man) in the South Central United States. *Aquaculture Research*, 2005, 36, 264-277.

Wurtz W.A., 2005. Low input shrimp farming in Kentucky. *World Aquaculture*, 38 (4): 44-49.

Index

.